WEATHER EXTREMES
in the WEST

Tye W. Parzybok

MOUNTAIN PRESS PUBLISHING COMPANY
Missoula, Montana • 2005

Photographs and illustrations by the author unless otherwise credited

Front cover photograph: Thunderstorm over Tucson, Arizona —Warren Faidley, Weatherstock

Library of Congress Cataloging-in-Publication Data
Parzybok, Tye W., 1970-
    Weather extremes in the West / Tye W. Parzybok.
        p. cm.
    Includes bibliographical references and index.
    ISBN 0-87842-473-3 (pbk. : alk. paper)
    1. West (U.S.)--Climate. I. Title.
    QC984.W38P37 2004
    551.6978--dc22

                                                    2004015440

PRINTED IN HONG KONG BY MANTEC PRODUCTION COMPANY

Mountain Press Publishing Company
P.O. Box 2399 • Missoula, Montana 59806
(406) 728-1900

For my wife, Jennifer Lynn, who gave me the strength, support, and time to write this book and to fulfill my dream; and to my boys, T. J. and Dylan, who mean more to me than a lifetime of extreme weather.

1. The Pacific Northwest's Coast, Western Valleys, and Mountains
2. The Cascades
3. The Pacific Northwest's Inland Empire and High Deserts
4. California's Coast, Western Valleys, and Southern Mountains
5. The Sierra Nevada
6. Desert Southwest
7. Snake River Plain and Great Basin
8. Northern Rocky Mountains
9. Central Rocky Mountains
10. Colorado Plateau and Southern Rocky Mountains
11. Southern Great Plains and Adjacent Foothills
12. Rocky Mountain Front and Adjacent Foothills
13. Northern Great Plains

Climate regions of the western United States covered in this book

Contents

Preface

Ever since weather first captivated me in elementary school, I dreamed of someday having the knowledge and background to make a contribution to the understanding of weather. So in 1995, at age twenty-five, I started work on this book. I knew that extreme, unique, and record-breaking weather has a way of capturing attention and influencing lives. Writing such a book appealed to me, and the growing weather library seemed to lack this resource.

Humans are immensely fascinated with weather, particularly the extremes, and the recent apparent increase in these phenomena has raised the public's awareness of, and interest in, such weather events. I say "apparent increase" because by definition extreme events occur infrequently. Furthermore, we have not collected scientifically sound data long enough to know if the frequency of weather extremes is increasing or even how often these extremes occur. This book does not attempt to address the complex debate over what might be causing the apparent increase in extreme weather. It simply focuses on the facts and details of the extreme events themselves and leaves the theory for another book.

The highly varied topography of the western United States gives rise to some of the world's most remarkable and hazardous weather. From tropical storms to blizzards, the West sees it all. And while you can get the daily weather forecast on the radio or TV, I hope this book will help you become weather-wise. You'll find here discussions of unique microclimates, storms, awe-inspiring weather records, famous natural disasters, and fascinating weather hazards in the West. In many cases the extreme phenomena help explain the hows and whys of normal weather and climate across the West. The extremes are, after all, the normal gone drastically amiss.

I wanted to fill this book with as many relevant weather facts as possible. As I came across interesting weather topics and facts, I wrote them down and later included them in this book. As I neared the end of the manuscript, I found it difficult to draw the line on the last fact or topic I would include. Please forgive me if I have left out a weather event significant to you.

I went to great lengths to ensure the accuracy of the facts, discussions, and dates in this book. However, extreme events can invite distorted views from those directly affected. Furthermore, these events push meteorological measuring equipment beyond its operating limits, forcing people to estimate such phenomena as wind speed, rainfall, snowfall, temperatures, and wave height. So even though I checked and cross-checked dates, data, and reports with other sources, this book may still contain inadvertent fiction or exaggerated reports.

In many ways we are all meteorologists. Every day we share our weather forecasts, observations, and stories through casual conversation. Most of us have experienced an extreme event—whether a blizzard, tornado, heat wave, or hurricane—and that event likely left a lasting impression on us. Even as a professional meteorologist and long-time weather enthusiast, I was touched by stories people told of violent weather's effects on them and their loved ones. Words lack the strength to describe the deadly weather events in this book; words describing the drama and tragedy of these events pale in comparison to experiencing them firsthand.

On a lighter note, I tried to mix in some of the West's unique—and not necessarily deadly—weather phenomena. The choices seem endless, so here again, I used my professional judgment to decide what to include and what to leave out. And again, I may have inadvertently left out unique weather topics that some readers will think I should have included.

Even as I write the last words of the manuscript, this book is to some degree outdated.

In one week in early May 2003, the National Weather Service (NWS) issued an astounding 4,867 warnings: 1,090 for tornadoes and 683 for flash flooding. On May 6 alone the NWS issued 908 warnings, the most ever issued in one day since record keeping began in 1986. The Santa Ana firestorm of October 2003 is ending as I finish writing. With the weather constantly changing, it throws more remarkable events at us every day. Just yesterday, here in Fort Collins, Colorado, we enjoyed warm temperatures in the mid-80s. Today, however, the temperatures hovered in the mid-20s, and we experienced freezing drizzle and light snow—a remarkable cool down of 60°F in one day! With weather's variability in mind, I invite you to send me your additions, forgotten events, or more accurate accounts of the events already discussed. Just like the weather, this book will change, but as written, it represents events that occurred through 2002, as well as a few that occurred in 2003 and 2004.

Acknowledgments

Although much of this book was a solo effort, I simply could not have accomplished it without the patience, help, support, and enthusiasm of a number of people. First and foremost, I thank my wife, Jennifer, who sacrificed a tremendous amount of time with me so that I could work toward this dream. I also thank my family for their constant support and interest.

I express great thanks to my reviewers and editors—Laura Braaten, George Taylor, Nolan Doesken, Peter Stoll, Gene Petrescu, Kathleen Ort, and especially James Lainsbury—for their invaluable comments, suggestions, and corrections.

Kind photographers gave me the honor of publishing many of the wonderful photographs in this book without a royalty. Thanks also to Christopher Davey for his help with illustrations.

I cannot thank Kathleen Ort of Mountain Press Publishing enough for accepting my manuscript. It was largely the work of the Mountain Press staff that made this book look as good as it does.

And thank you to all those who trusted me and provided me invaluable opportunities to become the professional meteorologist I am; in particular I owe special thanks to Jim Wirshborn (Mountain States Weather Services), Nolan Doesken (Colorado Climate Center), George Taylor (Oregon Climate Service), Chris Daly (Spatial Climate Analysis Service), Wayne Gibson (Oregon Climate Service), Ed Tomlinson (Applied Weather Associates), Lesley Julian (National Weather Service), and Geoff Bonnin (National Weather Service). Thanks again to everyone who helped this dream come true.

Introduction

The U.S. has sustained 42 weather-related disasters over the past 20 years in which overall damages and costs reached or exceeded $1 billion. Thirty-six of these disasters occurred during the 1988–1999 period with total damages exceeding $170 billion. Seven occurred during 1998 alone—the most for any year on record, though other years have recorded higher damage totals.

—National Climatic Data
Center, July 20, 1999

Although defining an extreme weather event is somewhat subjective, loss of life and property provide objective measures, along with the event's significance as it relates to weather records dating back one hundred years or more. In the United States alone, hurricanes, floods, and tornadoes account for an average $9.2 billion (1997 U.S. dollars) in damages every year, with general extreme weather causing $14 billion (2004 U.S. dollars) in losses a year.

The size of the United States and its particular geographical and societal conditions make it susceptible to influences associated with extreme weather. Extreme weather has always accounted for a great deal of death and destruction, and as populations continue to move into areas prone to dangerous weather—such as high mountains, coastlines, and deserts—we inherently become more vulnerable to the weather.

Since we have no systematic reporting method and no single repository for data regarding weather-related property loss and human death in the United States, it is difficult to determine accurately losses from weather hazards. *Storm Data and Unusual Weather Phenomena,* a monthly publication of the National Climatic Data Center—the nation's "scorekeeper" of severe weather events and their historical perspective—contains a chronological listing, by state, of extreme weather as reported by the National Weather Service. The publication also contains statistics on personal injuries and damage estimates, but evidence suggests that direct losses have been grossly underestimated in some cases. According to the Environmental and Societal Impacts Group at the National Center for Atmospheric Research (NCAR), weather-related losses to insured property in the United States have increased steadily since the mid-1990s. NCAR estimates that weather-related hazards killed more than 24,000 people and injured about 100,000 in the United States between 1975 and 1994; that equates to an average of 1,263 deaths per year.

Insurance companies report an exponential increase in disaster losses in the last twenty years and describe it as a "catastrophe trend." In the last twenty years alone this country reported $170 billion in loss of life and destruction from weather-related disasters. These increases in life and property losses stem from increasing vulnerability arising from societal changes, including

1

Aftermath of the 1900 Galveston hurricane, the worst natural disaster in terms of lives lost in U.S. history —Photo courtesy of the National Oceanic and Atmospheric Administration, Department of Commerce

growing populations in high-risk coastal and low-lying areas, more property subject to damage, and higher property values.

Since 1989, the United States has frequently entered periods in which losses from catastrophic natural disasters averaged $1 billion per week. Seven of the ten most costly disasters in U.S. history, based on dollar losses, occurred between 1988 and 1993. The United States isn't the only country experiencing damaging weather. According to the Worldwatch Institute, in 1998 alone, severe weather caused more than thirty thousand deaths and close to $90 billion in damage worldwide.

Introduction to Climate

Climate is the historical record and description of average daily and seasonal weather events over a long period—the average weather over decades. Meteorologists generally draw weather statistics from several decades, with three decades generally accepted as a long enough record to provide reliable averages. The concept of climate usually refers to the average or mean course of the

Texas Twister

The Galveston, Texas, hurricane of September 8–9, 1900, was the deadliest weather disaster in U.S. history. Winds to 120 miles per hour and a 20-foot storm surge accompanied the hurricane. The huge storm surge washed over the island, demolishing or carrying away buildings and drowning more than 6,000 people. The hurricane destroyed more than 3,600 houses, and damage totaled more than $30 million. The storm surge pushed the surf 300 feet inland from the former waterline. The hurricane claimed an additional 1,200 lives outside of Galveston.

weather. A young girl provided a colleague perhaps the best comparison of climate and weather. The youngster said, "Weather is what you get, and climate is what you're *supposed* to get."

The general climate of an area depends on a number of geographical factors including latitude, elevation, proximity to the ocean or other large water body, and mountains. Microclimates—localized climates that differ from the general climate of the area—depend on slope, aspect, surrounding topography, vegetation, irrigation, and proximity to urbanized areas, all factors that can affect temperatures and evaporation rates.

On a planetary scale, winds, temperature, and pressure govern Earth's climate. All three of these driving forces result from unequal solar heating of the earth. In short, then, the earth's almost unlimited variety of climates stems from variations in the amount of solar radiation striking the planet. The word *climate* comes from the Greek word *klima,* meaning "inclination," and it reflects the importance early scholars attributed to the sun's influence on Earth's weather patterns.

At tropical latitudes the sun's rays strike nearly perpendicular to the earth's surface. This delivers an intense amount of solar energy per unit area and is the chief reason for consistently warmer temperatures at tropical latitudes. The opposite is true at the poles, where the sun's solar energy strikes the earth at an angle, spreads over a larger area, and diffuses its energy. Since solar energy at the poles is far less intense per unit area than near the equator, the poles receive less heat.

Everybody knows the earth has four seasons. The primary reason we have seasons is the 23.5-degree angle of the earth's axis with respect to the plane of its orbit around the sun. In other words, the earth is tilted. As the earth orbits the sun, the northern hemisphere is at times oriented closer to or farther away from the sun. During the northern hemisphere's summer, the earth is oriented such that the northern hemisphere is closer to the sun. At this time of year the sun rises higher in the sky, stays above the horizon longer, and strikes the ground more directly with its rays. The increase in direct sunlight and daylight hours leads to more surface heating and summerlike conditions.

During winter, the northern hemisphere is oriented away from the sun. At this time of year,

The Nation's Costliest Weather Event

Hurricane Andrew (August 16–28, 1992) was the costliest weather disaster in U.S. history. The Category 4 hurricane hit Florida and Louisiana with high winds that damaged or destroyed more than 125,000 homes, and damages totaled approximately $27 billion. The storm killed sixty-one people, but if the storm had been slower and wetter, or if its path had brought it close to Miami, it would have caused far more widespread death and devastation.

the sun rises low in the sky, remains above the horizon for a shorter period of time, and strikes the ground less directly with its rays. This results in less surface heating and colder temperatures. Spring and fall are transitional seasons.

Contrary to what you might think, air is mainly heated from below—that is, the earth heats the air rather than the air heating the earth. As the sun's radiation hits the earth, it warms the surface, which heats the air above it. Since warm air is less dense than the overlying cool air, the warm air rises. The rising warm air displaces cooler air aloft, which then sinks. This cooler air is then heated in the lower levels of the atmosphere and then rises as well. In this way a simple atmospheric circulation, known as *convection,* is born.

This circulation pattern occurs along latitudinal belts around the earth. The zones of circulation constantly distribute excess heat from the equator toward the cold polar regions. Without this air and heat circulation, the equator would continue to warm and the poles would continue to cool. Scientists now recognize global climate zones as largely based on the earth's general circulation.

Along the equator, with its maximum heating throughout the year, the air tends to rise, forming clouds, rain, and a band of low pressure. Some of the world's wettest (and warmest) places

Hurricane Andrew's storm surge and heavy rains stacked these cars in Miami, Florida. —Photo courtesy of the National Oceanic and Atmospheric Administration/Department of Commerce

Mosaic of four satellite images, taken at approximately twenty-four-hour intervals starting at 2:19 PM eastern standard time on August 22, 1992, showing the progression of Hurricane Andrew as it cut a deadly and disastrous path across Florida and Louisiana —Image courtesy of National Oceanic and Atmospheric Administration, National Environmental Satellite, Data, and Information Service

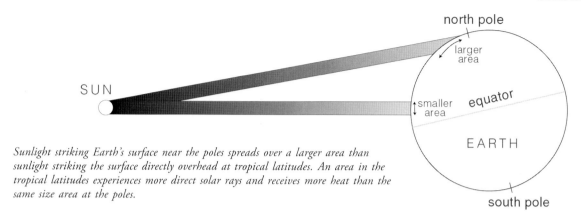

Sunlight striking Earth's surface near the poles spreads over a larger area than sunlight striking the surface directly overhead at tropical latitudes. An area in the tropical latitudes experiences more direct solar rays and receives more heat than the same size area at the poles.

sit along the equatorial belt of low pressure, also known as the Intertropical Convergence Zone or ITCZ.

Around 30 degrees north and south latitude the air that rose at the equator, now cooler, begins its return to the surface. This induces a band of sinking air characterized by high pressure, dry air, and limited precipitation—a circulation pattern called the *Hadley cell*. Most of the world's deserts are located along these high-pressure belts.

The circulation pattern between about 30 degrees north or south and 60 degrees north or south latitudes is called the *Ferrel cell*, or sometimes the *mid-latitude cell*. The warmer mid-latitude portion of the Ferrel cell meets the colder polar cell at the polar front—the location of the important polar jet stream. The Ferrel cell rotates warmer air from the mid-latitudes northward to the polar front, where it overruns the colder, denser polar air. Consequently, the overrunning air is lifted and cooled. The cooled air aloft then sinks and drifts toward the equator to fill the void the warm surface air created when it traveled north. This completes the Ferrel cell circulation.

The earth's rotation causes moving particles, including wind, to be deflected by the *Coriolis force*. In the northern hemisphere, the deflection is to the right, and in the southern hemisphere it is to the left. Air moving north in the Ferrel

cell circulation is deflected eastward. That deflection, known as the *Coriolis effect*, transforms the airflow into a prevailing westerly wind known as the *westerlies*. Existing in the mid-latitudes—generally between 30 and 60 degrees latitude—westerlies cause storms to move generally from west to east across the United States. Similarly, air associated with the southward movement of air in the Hadley cell is deflected westward and results in northeasterly winds between the equator and 30 degrees north. These winds are known as the *northeast trade winds*.

Which Way Does the Wind Blow?

Which direction does a westerly wind flow? Many people might assume that a westerly wind flows to the west, but the opposite is true. Westerly winds come out of the west, northerlies out of the north, easterlies out of the east, and southerlies out of the south.

At about 60 degrees north and south latitude, the thermal gradient that develops between the cooler high latitudes and the warmer subtropics causes the polar jet stream—called a *frontal jet stream*—to form. It received its name because it develops where two very different air masses converge. The dramatic change in temperature (thermal gradient) along this front produces dramatic changes in atmospheric pressure, setting

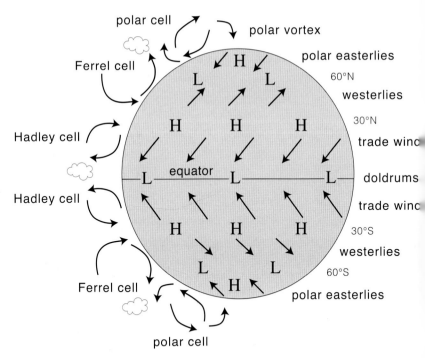

Simplified global three-cell surface and upper-air circulation patterns. This figure shows what the earth's global wind pattern would be like if water covered the earth uniformly and surface pressure was distributed equally. The northward flow of air from the Hadley cell is deflected east by the Coriolis effect, causing a zone of northeasterly winds known as the northeast trade winds. The Coriolis force causes the opposite to occur in the southern hemisphere. These winds intersect near the equator, where the net flow of wind nears zero, causing an area of relatively light surface winds called the doldrums. In the northern hemisphere, the Ferrel cell drives air northward. The air is deflected eastward, resulting in the important mid-latitude westerlies. Further north are the polar easterlies, which are caused as the Coriolis force deflects the polar cell circulation. The earth's landmasses and topography disrupt these circulations in complex ways, but this general flow persists.

up a steep pressure gradient and creating intense wind speed. The polar jet stream does not flow in a straight line. It meanders north and south through the year, as sinuous as a snake at times as it writhes across North America. In general, the jet exists near 45 degrees north latitude in winter and 60 degrees north in summer.

The earth's other major jet stream, called the *subtropical jet stream,* is located between the Ferrel and Hadley cells. It usually meanders between 30 and 40 degrees north latitude. Unlike the polar jet stream, which forms because of a large-scale thermal gradient, the subtropical jet is formed by the conservation of angular momentum. As the earth rotates, everything on the planet, including parcels of air, travels the circumference of the globe in twenty-four hours. Parcels nearest the poles don't travel as far or at all, and thus don't travel as fast. Parcels near the equator, however, travel over 1,000 miles per hour to complete their daily circulation around the globe. As an air parcel rises from the surface

of the equator (part of the Hadley cell), it moves northward. The air parcel was traveling at 1,000 miles per hour, as was the ground at the equator, but as the air moves northward, its speed increases relative to the speed of the ground at higher latitudes. This is called the *conservation of angular momentum.* This increase in speed to the east results in a strong westerly wind—the subtropical jet stream.

The polar and subtropical jet streams, strong rivers of air that typically circle the earth at altitudes of 30,000 to 35,000 feet (near the tropopause), greatly affect the weather at the surface. In reality, jet streams are not composed of a solid flow of air moving at the same speed, but of winds traveling at many different velocities. Jet streaks, also called *jet maximums* or *jet maxes,* are segments of the main flow that have higher wind velocities. They are caused by strong temperature gradients that develop over short horizontal distances in the lower troposphere. Jet streaks are important because they cause wind shear, which can lead to perturbations in the jet stream. Perturbations in jet stream flows can grow and develop into powerful storms, which then result in very strong warm and cold frontal boundaries that push into the United States.

The winds within the core of the polar jet stream can reach speeds of 200 miles per hour; the winds in the subtropical are usually slightly less. During winter, when temperature contrasts between the north pole and the mid-latitudes peak, the polar jet stream blows the strongest. For instance, the average January temperature difference between Fairbanks, Alaska, and Denver, Colorado, is nearly 40°F, −9.7°F and 29.2°F respectively. During July, however, the average temperature difference between these two cities only ranges 11°F (62.4°F versus 73.4°F), resulting in a much weaker temperature gradient and a weaker jet stream. The seasonal temperature shift and subsequent weakening of the polar jet stream largely explains why the western United States does not see large-scale storm systems in the summer. Summer's smaller-scale weather dynamics instead produce regional and localized storms, such as thunderstorms and rainstorms. Jet streams follow paths called *storm tracks,* because storms tend to form along and follow jet streams.

Winds blow at all levels of the troposphere, usually at different speeds and in different directions, but with increasing altitude the winds blow at a more uniform speed and direction. Overall they follow the general atmospheric flow around the earth. The layers of wind closer to the earth are influenced by the earth's topography and its differential heating (uneven heating rates across the earth's surface), which cause perturbations in the overall global flow. The winds that most influence daily weather exist at or below the jet stream level. The upper-level wind characteristics of speed and direction—halfway through the atmosphere, at about 18,000 feet above sea level or at 500 millibars of pressure—are important for forecasting. They provide a good indication of mid-atmospheric conditions—the conditions that have the most impact on the weather we witness on the ground.

Introduction to Weather

Weather is the state of the atmosphere at a specific time as described by variations in cloudiness, humidity, precipitation, temperature, visibility, and wind. The interaction of five factors causes weather: atmosphere, Earth's rotation, liquid water, sunlight, and topography. Weather can wear many different faces. And unlike climate, the weather can change in minutes or even seconds. It is a dynamic force, and we are only beginning to understand how it works.

The weather we experience takes place within a relatively thin slice of air surrounding the earth called the *atmosphere.* Think of the atmosphere as a constantly flowing river. This 10-mile-thick cocoon of gases separates Earth from the extreme and inhospitable climate of space. Oxygen, carbon dioxide, nitrogen, and water vapor make up 99 percent of the atmosphere; trace amounts of about fifty other gases make up the remaining 1 percent. The atmosphere consists of several layers: troposphere, stratosphere, mesosphere, ionosphere, and exosphere. Closest to Earth is the troposphere. Most of the clouds you see in the sky exist in the troposphere, and this is the layer of the atmosphere we associate with weather.

The properties of air—temperature, weight, and movement—make up some of the fundamental driving forces of weather. Consider an air parcel, that is, a sample of air we think of as homogeneous. The pressure, temperature, and humidity in this parcel are the same throughout. When the sun heats the air within this parcel, the gas molecules begin moving more rapidly, hitting the sides of the parcel more frequently and with more force, thus increasing the pressure in the parcel and forcing it to expand. In a cooling parcel of air the molecules move slower and the parcel

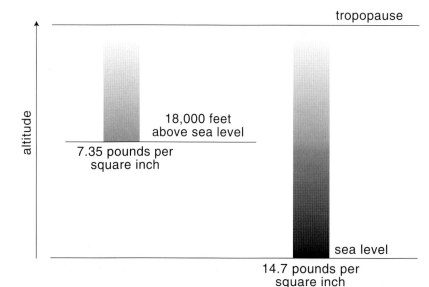

tropopause

altitude

18,000 feet
above sea level

7.35 pounds per
square inch

sea level

14.7 pounds per
square inch

Atmospheric pressure decreases rapidly with height. At sea level a square-inch column of air exerts 14.7 pounds of pressure on the earth's surface. Halfway up into the atmosphere—at an altitude of about 18,000 feet or 500 millibars—the pressure equals half the sea level pressure. This means that at 3.4 miles above the surface, we are above half of all the molecules in the atmosphere. The top of the troposphere, called the tropopause, varies in height with season (higher in summer) and latitude (higher near the equator), but in the mid-latitudes, its height averages about 7 miles above sea level.

of air contracts. Warming air expands and cooling air contracts. Also, cooler parcels of air are more dense than warm parcels since their molecules are packed more tightly; that's why cooler parcels of air tend to sink in the atmosphere and are often associated with high-pressure systems.

The sheer weight of air plays an important role in creating weather. Take for instance a 1-inch-by-1-inch square parcel of air above the earth's surface. At sea level, the average weight of the air in that parcel of air is 14.7 pounds. That means the column of air molecules between the ground and the top of the troposphere—typically 10 miles above the surface—exerts a pressure of 14.7 pounds per square inch at sea level, which is the same as 1013.25 millibars. (A millibar is the metric unit meteorologists use to measure pressure.) At elevations higher than sea level, there are fewer air molecules above 1 square inch of the earth's surface and the troposphere, so the air exerts less weight and, therefore, the atmospheric pressure is less. For instance, air molecules exert a pressure of about 4.4 pounds per square inch or about 300 millibars on the peak of Mount Everest, elevation 29,035 feet. Atmospheric pressure, then, equals the weight of a column of air per unit area.

Meteorologists measure atmospheric pressure with an instrument called a *barometer,* which is why we refer to atmospheric pressure as *barometric pressure.* One of the most common weather maps forecasters use is a map of isobars, which are lines connecting points of equal pressure. These weather maps resemble topographic maps.

Like waves in a river, air flows in the upper atmosphere in longwave troughs and ridges. Each low-pressure trough is balanced by adjacent, upstream and downstream, high-pressure ridges. Complex oceanic and atmospheric forces prevent the longwave pattern from changing drastically from day to day; only over several days will this large-scale pattern change, and in some cases the pattern can remain relatively constant for up to two months.

Both jet streams bulge north of ridges and south of troughs. On a smaller scale at the surface, relatively smaller waves, or "ripples," move along the jet stream as storms (cyclones). Several factors cause these waves, including uneven surface heating and wind shear caused by jet stream maximums. As with the large-scale longwave pattern, these storms have strong high-pressure areas (anticyclones) ahead of and behind them. The interaction between large-scale systems (troughs, ridges, and jet streams), smaller-scale systems (surface cyclones and anticyclones), and global forces (Coriolis effect and pressure gradient force) influences the movement of weather systems.

Variations, or gradients, in air pressure—caused by differences in air temperature—develop between cyclones and anticyclones, leading to the development of surface winds. In general, winds blow from regions of high pressure to areas of low pressure. Imagine a filled balloon. The air inside the balloon has relatively higher pressure than the air surrounding it. When you pop the balloon, the air travels through the hole, from the region of higher pressure to the surrounding lower pressure of the air mass surrounding it.

The speed of wind is proportional to the pressure gradient. The pressure gradient force is equal to the Coriolis force in the upper atmosphere, which results in a balanced flow, or as meteorologists call it, a *geostrophic flow*. In other words, the wind blows parallel to isobars that weave around high- and low-pressure systems without resulting in a net increase or decrease of air at the pressure centers.

The frictional effects of topography, however, retard winds at the surface. This causes an imbalance between the Coriolis and pressure gradient forces. As a result, the pressure gradient force dominates the flow of wind in the lower atmosphere, and winds pile up in low-pressure systems. In general, this is how cyclones form; anticyclones form in response to cyclones.

The piling up of winds induces a rising motion—since the air has nowhere else to go—and forces the air at the surface to rise, cool, condense, and eventually turn into clouds. Cyclones are usually associated with clouds and precipitation, and they always flow in a counterclockwise direction in the northern hemisphere. Anticyclones—areas of high pressure that circulate in a clockwise direction in the northern hemisphere—have the opposite

How winds behave in high- and low-pressure systems in the upper atmosphere and near the earth's surface. (a) The Coriolis force and pressure gradient force are equal in the upper atmosphere. Without friction to retard its movement, wind blows freely around high- and low-pressure systems. (b) Friction slows winds near the surface, weakening the Coriolis force, which depends on speed. Because the pressure gradient force remains unchanged, air flows into low-pressure areas and out of high-pressure areas.

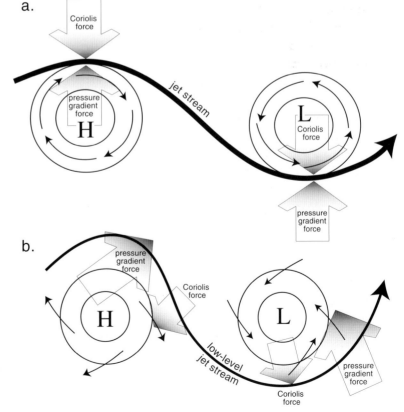

characteristics: sinking motion, dry air, and clear skies. It's important to remember that these systems "hold" each other together. Without anticyclones, cyclones would spin themselves apart.

An easy way to remember the important aspects of these two circulation systems is what meteorologists call the *right-hand rule.* If you hold your right hand out with your thumb up, your fingers curl in a counterclockwise (cyclonic) direction while your thumb points in the direction of a cyclone's airflow (upward). If you flip your hand over so your thumb points down, your thumb and fingers represent the airflow in an anticyclonic system (high pressure).

As I mentioned before, wind—caused by the interaction of high-pressure and low-pressure systems—is nature's way of transporting heat and moisture around the earth in an attempt to reach equilibrium. Strong winds are associated with strong pressure gradients, light winds with weak pressure gradients. You can easily discern strong pressure gradients on a weather map by looking for the telltale "tight packing" of isobars. Maps of isobars are useful for locating areas of high and low pressure, which correspond to the positions of surface cyclones and anticyclones. On the ground, Buys Ballot's law provides an easy way to figure out where you are in relation to cyclones and anticyclones. It says that in the northern hemisphere, if you stand with your back to the wind, the low pressure is on your left and the high pressure is on your right.

Besides pressure systems, wind, sun, and Earth's rotation, the amount of available liquid water strongly influences weather. As water evaporates from oceans and other large water bodies, it rises into the atmosphere. As it rises, atmospheric pressure decreases rapidly with altitude and exerts less and less pressure on the air, allowing it to expand and cool. As the air mass cools, the molecular activity of its water vapor slows until the molecules condense to a liquid, and a cloud is born. Some of the water droplets collect on

small particles called *condensation nuclei,* which can be dust, salt, volcanic ash, or smoke particles. Eventually this becomes precipitation. The speed at which the water molecules move around in the air determines whether the water molecules will be gas (fast movement), liquid, or solid (slow movement). So as air cools, water molecules slow down and begin to change from gas (water vapor) to liquid (rain, clouds, or fog) or solid (snow or hail). The opposite of this process, evaporation, occurs when an air mass is heated and liquid moisture becomes a gas.

The atmosphere has three key mechanisms it uses to force air to rise, cool, condense, and consequently precipitate: convection (air lifted by rising warm air), convergence (air piled up in a region and forced to rise), and topography (orographic lifting). Water vapor can also be cooled and converted into water droplets without lifting. This process, called *advection,* occurs when warm, moist air blows horizontally across a much colder surface, which causes the air to cool below its dew point. The dew point is the temperature at which an air mass is saturated with moisture. Once the air mass is saturated, condensation occurs.

There are many different types of clouds. Cirrus clouds form high in the troposphere and consist primarily of ice crystals. Strong winds in the upper troposphere make cirrus clouds look wispy and thin. Stratus clouds form in lower parts of the troposphere, consist of water droplets, and cover most of the sky with a uniform gray similar to a fog. Cumulonimbus clouds are tall, dense, cauliflower-like clouds with anvil-shaped tops. These clouds signal thunderstorms and also spawn tornadoes and other violent weather.

Atmospheric instability is necessary not only for clouds to form, but also for storms. Instability occurs when a relatively warm air parcel rises into cooler air. Technically, the amount of instability is determined by comparing the environmental lapse rate—the temperature change with increasing height—to the moist and/or dry adiabatic

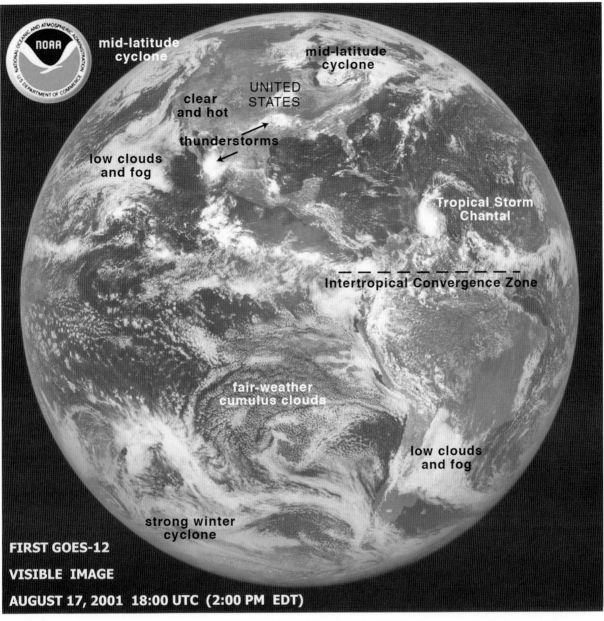

mid-latitude cyclone

mid-latitude cyclone

UNITED STATES

clear and hot

thunderstorms

low clouds and fog

Tropical Storm Chantal

Intertropical Convergence Zone

fair-weather cumulus clouds

low clouds and fog

strong winter cyclone

FIRST GOES-12

VISIBLE IMAGE

AUGUST 17, 2001 18:00 UTC (2:00 PM EDT)

Full-disk visible satellite image taken from the Geostationary Operational Environmental Satellite (GOES) 12 on August 17, 2001, at 2:00 PM (eastern daylight time). This image shows a typical day of weather on Earth. Tropical Storm Chantal, which had sustained winds of 40 miles per hour, strengthened as it moved across the eastern Caribbean Sea. A cluster of tropical thunderstorms not associated with a tropical storm or depression rumbled over the Gulf of California. Two strong mid-latitude cyclones spun in the northern hemisphere, one across the northern Pacific Ocean and the other over the Great Lakes. A heat low kept the U.S. desert Southwest hot and dry while thunderstorms hammered the Mississippi Valley. Areas of low clouds and fog covered portions of south-central South America and the east-central northern Pacific Ocean. Across the southern hemisphere, where it was winter, large storms and a great deal of cloudiness—including fair-weather cumulus clouds—prevailed across the southern Pacific Ocean. On a larger scale, a band of clouds and thunderstorms along and near the equator marked the Intertropical Convergence Zone or ITCZ. —Image courtesy of the National Oceanic and Atmospheric Administration/Department of Commerce; annotated by author

Incredibly strong tornadic winds rammed this 33 rpm record into a downed telephone pole. —Photo courtesy of the National Oceanic and Atmospheric Administration/ Department of Commerce

rates. When the environmental lapse rate is greater than the dry adiabatic or moist adiabatic rate, the atmosphere is considered unstable.

The dry and moist adiabatic rates refer to the rate of temperature change in rising or descending parcels of air. The rate for dry (unsaturated) parcels is fixed: 5.5°F per 1,000 feet. The rate for moist parcels (saturated) varies. A common rate is 3.3°F per 1,000 feet. As a dry or partially saturated parcel rises, it cools. Eventually it cools to its dew point—when it is saturated and can no longer hold all its moisture—and from then on the parcel cools at the moist adiabatic rate. The altitude at which a rising air mass reaches its dew point is called the *level of condensation;* this occurs when clouds and precipitation start to form, eventually creating a storm.

Hurricanes, thunderstorms, hailstorms, tornadoes, blizzards—the list could go on. They fascinate and scare us. A storm, as defined by the Weather Channel, is "an individual low pressure disturbance, complete with winds, clouds, and precipitation. The name is associated with destructive or unpleasant weather." By-products of the earth's global transfer of energy, for many weather nuts storms are entertainment. The West, as you will read, experiences all of these storms almost on a yearly basis, and they cost the United States billions of dollars. Some of them even kill.

Climate and Weather in the West

The United States west of the hundredth meridian, which I refer to as "the West" in this book, experiences an endless and fascinating variety of weather influenced by its rugged topography and the vast Pacific Ocean. With the exception of the Pacific Northwest, where eastward moving cyclones drop abundant moisture, the West is relatively dry compared to the rest of the country. Even so, the West can claim the country's driest and wettest places. If not obstructed by dust, the dry air across the western United States provides excellent visibility, and it frequently exceeds 100

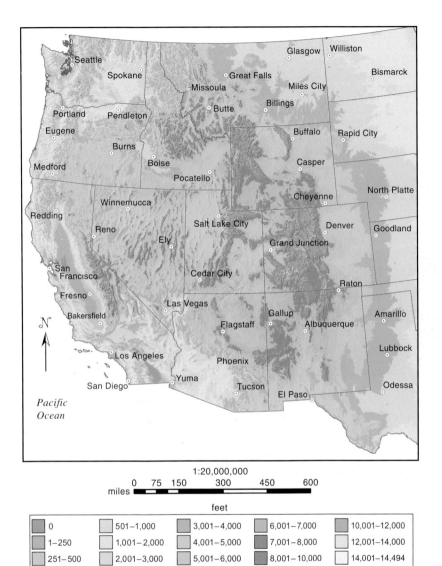

Shaded relief map of the West

Pacific Ocean

1:20,000,000

| 0 | 75 | 150 | 300 | 450 | 600 |
miles

feet

0	501–1,000	3,001–4,000	6,001–7,000	10,001–12,000
1–250	1,001–2,000	4,001–5,000	7,001–8,000	12,001–14,000
251–500	2,001–3,000	5,001–6,000	8,001–10,000	14,001–14,494

miles, a rarity in the more humid eastern United States. The West also experiences extreme variations in temperature and holds records for the all-time highest and lowest daily temperatures in the contiguous United States.

The main source of moisture in the West is the Pacific Ocean; however, the Gulf of Mexico provides moisture to the Rocky Mountain Front and adjacent Great Plains, while the tropical Pacific provides moisture to the semiarid Southwest.

Three main storm tracks cross the West. During winter, storms can track almost anywhere across the West, but they favor a southern path across California, Arizona, and New Mexico. In summer and fall, storms tend to track farther north across Washington, northern Idaho, and Montana. In spring storms track predominantly through the middle of the West, with a favored path across central California to the Four Corners region and then east.

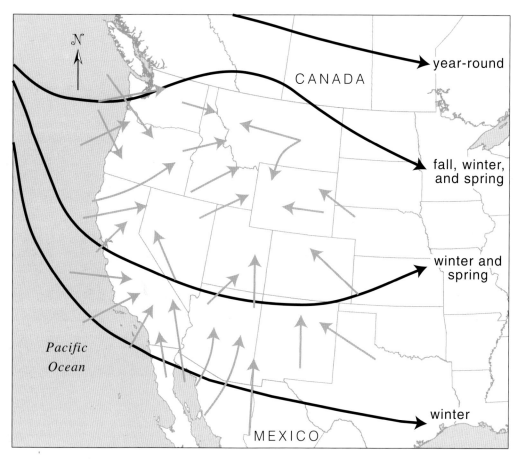

Major storm tracks across the West (black lines) and principle paths of moisture inflow from storms producing large amounts of precipitation (gray lines). Adapted from NOAA Atlas 2.

Regardless of the track, the region's rough terrain typically weakens storms traversing the West. Storms re-intensify, however, as they cross the Continental Divide and expand vertically in the deeper atmosphere over the plains. Even after a storm has moved east of the Continental Divide, its often explosive redevelopment and expansion over the Great Plains provides the Rocky Mountain Front and western Great Plains with a variety of severe weather; in summer it can provide the catalyst for hail, tornadoes, high winds, or flooding rains, while in winter or spring it can bring a severe winter storm.

The West's complex topography is largely responsible for its abundance of microclimates. From local winds to uniquely cold or hot locations, remarkable variations can exist from place to place—whether a few miles or a few hundred miles apart. One of the most dramatic rain shadows—a microclimate that keeps regions extremely dry—in the world is located in the mountains of western California, Oregon, and Washington. The West, particularly along the eastern slopes of the Rocky Mountains, endures the strongest nonhurricane and nontornadic wind gusts in the country, at over 100 miles per hour.

Average annual precipitation map (based on data collected between 1961 and 1990) created using Oregon State University's Spatial Climate Analysis Service PRISM (Parameter-elevation Regressions on Independent Slopes Model)—an analytical tool using point data, a digital elevation model, and other spatial data sets. PRISM generates gridded estimates of monthly, yearly, and event-based climatic parameters, such as precipitation, temperature, and dew point. Dr. Christopher Daly at Oregon State University developed and maintains PRISM. —Oregon State University/Oregon Climate Service

1:20,000,000

| | | | | | |
| miles | 0 | 75 | 150 | 300 | 450 | 600 |

inches

< 5	20.01–25	40.01–45	70.01–80	120.01–140
5.01–10	25.01–30	45.01–50	80.01–90	140.01–160
10.01–15	30.01–35	50.01–60	90.01–100	160.01–180
15.01–20	35.01–40	60.01–70	100.01–120	> 180

Microclimates like these make western weather particularly difficult to predict and extremely interesting.

Although the West claims many of the United States' extreme weather records, most of the time it enjoys sunny and tranquil weather. In fact, portions of the desert Southwest are the sunniest places in the United States, even sunnier than Florida—the Sunshine State!

How to Use This Book

This book is divided into thirteen chapters. Each chapter represents a region with similar weather, climate, and topography. I largely define the regions based on average annual precipitation, including melted snow. That said, in the context of region, some vignettes for particular places may seem out of place. Also, because a lot of weather phenomena affect broad areas, a weather discussion in one chapter may discuss several regions.

At the beginning of each chapter, I include an overview of a region's weather and climate, as well as a climate graph from a weather station representative of the region and climate characteristics. The climate characteristics include the Köppen climate classification of a particular region, a widely used classification system of world climates based on annual and monthly temperature and precipitation averages. The classification provides a general description of a region's climate. It was devised by the German scientist Waldimir

Köppen (1846–1940), who initially published the system in 1918.

The Köppen classification system distinguishes five major climatic types:

Tropical moist: all months have an average annual temperature above 64°F

Dry climates: water deficit (potential evapo-transporation exceeds precipitation) most months of the year

Moist mid-latitude climates with mild winters: warm to hot summers, mild winters, average temperature in the coldest month below 64°F but above 27°F

Mean annual snowfall of the western United States (based on data collected between 1961 and 1990) created using Oregon State University's Spatial Climate Analysis Service PRISM

1:20,000,000

miles 0 75 150   300   450   600

inches

0–5	20.1–30	50.1–75	150.1–200	400.1–500
5.1–10	30.1–40	75.1–100	200.1–300	500.1–750
10.1–20	40.1–50	100.1–150	300.1–400	> 750

Moist mid-latitude climates with severe winters: warm summers, cold winters, average temperature in the coldest month below 27°F

Polar climates: extremely cold winters and summers, average temperature of the warmest month below 50°F

Each of the main groups is further subdivided to describe regional climatic characteristics. The West falls into three of these main categories: moist mid-latitude climates with mild winters, dry climates including desert or steppe, and Highland climates. Mountainous terrain, where rapid changes in elevation result in sharp temperature and precipitation variations, makes delineating the general climate impossible. That is why we designate these regions as Highland climates.

Each chapter contains several extreme, historic, and/or unique weather- and climate-related vignettes that occurred or still occur in a particular region. I intend each vignette to be a freestanding discussion of how and why a particular weather- or climate-related phenomenon occurred or occurs. Because of this, I repeat some important terms and concepts in several chapters. The vignettes in this book represent a collection—but not all—of the West's most interesting, extreme, unique, and historic weather- and climate-related phenomena. A complete list of references and sources of information follows at the end of the book.

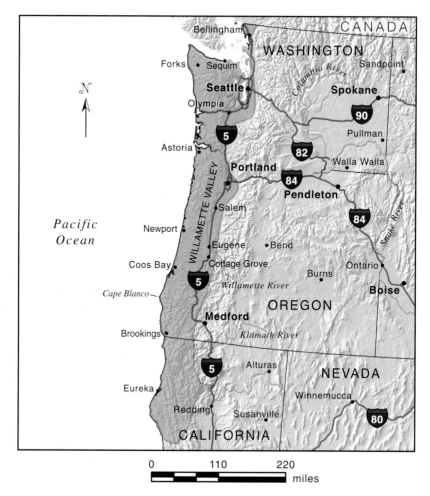

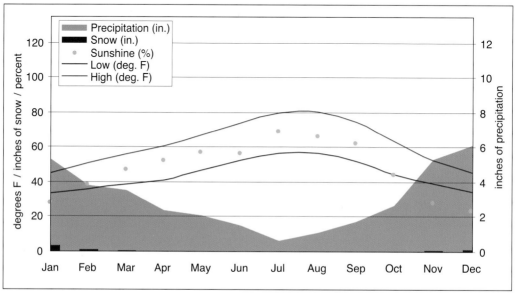

Climate of Portland, Oregon

The Pacific Northwest's Coast, Western Valleys, and Mountains

Elevation: sea level to 8,000 feet

Natural vegetation: needleleaf evergreen forests

Köppen climate classification:
Mediterranean, with moist, mild winters, and dry, cool summers

Most dominant climate feature: prolonged periods of light rain and drizzle during the winter

By January and February, the wet season in the Pacific Northwest is well underway as a continual barrage of Pacific storms slams the region. Typically, precipitation ranges from 12 to more than 40 inches during these months. The heaviest amounts fall in the Coast Ranges, Olympic Mountains, and west-facing foothills of the Cascades, where extraordinary snowpacks develop. During an average year, Seattle, Washington, and Portland, Oregon, receive precipitation on two of every three days. Average daily temperatures drop to their coolest levels of the year: 30s inland and 40s near the coast. The cool temperatures and abundant moisture commonly provide snow, even to the lowest elevations when arctic air enters the region. Low-elevation snowstorms usually do not last long since milder Pacific air returns when winds turn westerly.

Unrelenting rains persist into March and April, and most areas receive between 8 and 20 inches of precipitation. Average daily temperatures moderate into the 40s and 50s. Most inland areas

experience their last spring freeze in late April, while coastal areas are frost free by late March.

By May and June, the persistent winter rains that began on a regular basis in November start to ease. Before the rain ends in late June, 4 to locally 8 inches of rain typically falls. Average daily temperatures slowly climb into the comfortable 60s.

Summer is in full swing by July, with peak heating and dryness occurring in August. The predominant wind flow shifts from a southerly to a northwesterly direction, ushering in cooled air from the cold waters of the north Pacific and providing the region natural air conditioning. By July some residents may even wish for rain since this region rarely receives appreciable precipitation midsummer. Occasional showers and thunderstorms bring welcome precipitation to the higher elevations and sometimes to the lowlands. The Olympics can receive up to 3 inches of precipitation in July and August; elsewhere an inch or less is normal. While intense heat typically grips the rest of the country, the Pacific Northwest enjoys ocean-moderated temperatures in the 70s and 80s during the day and overnight lows in the 40s and 50s. Occasionally a heat wave will drive temperatures into the low 100s inland, but the coast rarely gets that warm. Fog and cool temperatures commonly shroud the coastal areas during the summer.

September and most of October bring perhaps the best weather to the Pacific Northwest. Nights are cool, with temperatures in the 40s and

50s, and days are mild, with temperatures in the 60s and 70s; skies are mostly clear. Coastal areas have the occasional exception, as cool marine air can produce fog and low clouds that hold temperatures in the 40s and 50s. Most areas feel their first freeze by late October. Although September is dry, October can turn wet; most areas receive 2 to locally 6 inches.

The wet season typically starts in earnest by November. Precipitation totals during November and December range from 10 to 20 inches. Temperatures cool into the 40s and 50s during the day, while morning lows dip into the 30s inland and 40s nearest the coast. Snow can fall in the lowest elevations, but for the most part, snow accumulation is confined to the higher elevations. Ice storms, however, can threaten the inland lowlands, especially near the Columbia River Gorge. Powerful Pacific-born storms can lash the coasts with winds of more than 80 miles per hour and sometimes more than 100 miles per hour. Winds throughout the winter blow predominantly out of the south.

The West Coast's Deadliest Twister: Washington and Oregon

April 5, 1972, started as a typical spring day in the northern Willamette Valley—a few clouds, warm temperatures, and light wind. By noon, however, ominous thunderheads began to build. By 2:00 PM rare and severe thunderstorms began moving across western Oregon. Around 2:45 PM, a funnel cloud formed over Portland, and at 2:50 PM the funnel cloud touched down near Portland at the southern shore of the Columbia River, damaging boats, moorages, and houses on Marine Drive. The strengthening twister then crossed the Columbia River, drawing up water along the way and gained the strength of an F3 tornado (see F-scale table in *Tornado Alley* in *chapter 11*). It tore a 9-mile-long path of destruction through Vancouver, Washington, severely damaging a grocery store, bowling alley, and grade school near the site of the present-day

Vancouver Mall. The tornado killed six people, injured three hundred, and inflicted $50 million ($215.5 million in 2002 dollars) in damage. Later that day, another F3 tornado touched down west of Spokane, and an F2 touched down in rural Stevens County, both by-products of the same large-scale storm system passing across the Pacific Northwest. The Vancouver tornado stands as the most deadly tornado to strike Washington, and the National Weather Service ranks it the seventh most significant weather event in Washington during the twentieth century.

Deadly Oregon Tornado

The greatest loss of life due to a tornado in Oregon occurred near Jordan Butte, northwest of Lexington in Morrow County. On the afternoon of June 14, 1888, the tornado swept through Lexington, Sand Hollow, and Pine City, killing six, injuring four, and destroying thirty buildings including two schools.

Thunderstorms and tornadoes rarely occur in western Oregon and western Washington. The region experiences an average of three to six thunderstorm days per year; a storm in this region lasts an average of fifty minutes—among the shortest in the country. Oregon and Washington combined average only one tornado per year. During 1997, however, Washington reported a record fourteen tornadoes. They were weak tornadoes, however: seven registered as F0s and the other seven as F1s.

Though rare, tornadoes and funnel clouds can strike west of the Cascades any time of year, unlike regions of the Great Plains and Midwest that have a fairly predictable tornado season running from April through July. Generally the by-products of highly energized winter storms that slam the Pacific Coast, these storms have unstable air masses (much colder air aloft than near the surface) and strong upper-level winds.

They create marginally favorable conditions for weak tornadoes, funnel clouds, waterspouts, and cold air funnels. The cold air aloft forces the relatively warmer air near the surface to rise rapidly into the atmosphere. This strong upward motion, coupled with strong winds from various directions throughout the atmosphere, causes columns of rising air to rotate into weak funnel clouds, which are classified as tornadoes once they touch the ground. Unlike the massive tornadoes of the Great Plains, which form when large thunderstorms release built-up instability, the weaker rotating columns of air that occur in the Pacific Northwest result from quick, localized releases of instability. The Great Plains also has more favorable wind shear conditions—winds blowing at different speeds and directions at different altitudes—for large twisters.

Wallowa County Twister

Around 4:00 PM on June 11, 1968, a rare tornado struck Wallowa County, Oregon, in a mountainous, uninhabited timbered area. The tornado destroyed approximately 1,800 acres of prime timber and badly damaged an additional 1,200 acres (an estimated 40 million board feet of lumber). Lasting about five minutes, the storm cut a ground path about 9 miles long and nearly 2 miles wide and delivered golf ball–sized hail. The National Weather Service ranks the storm as Oregon's ninth top weather event of the twentieth century.

While tornadoes are rare in Washington and Oregon, "cold core" funnel clouds and waterspouts are not. Like tornadoes, "cold core" funnel clouds and waterspouts form rotating vertical columns of air from updrafts and low-level wind shear. Not considered tornadoes, these funnel clouds are usually short-lived circulations that cause minimal damage. They develop from relatively small showers or mild thunderstorms with weak updrafts, which are associated with strong storm systems that have very cold upper-atmospheric temperatures.

The most damaging tornadoes in the Pacific Northwest occur in the spring and summer in association with intense localized thunderstorms. Throughout most of the year though, the cold sea-surface temperatures off the coast cool the surface

Washington's Rare Tornado

The strongest twister to hit the Puget Sound touched down near the Boeing Company in Kent, Washington, at 4:15 PM on December 12, 1969. This tornadic thunderstorm was associated with a vigorous low-pressure area sweeping across the Evergreen State. Fortunately, the rare twister did not claim any lives, but it did injure one person and cause about $460,000 ($2.3 million in 2003 dollars) in damage.

Photographed on September 10, 1969, from an aircraft off the Florida Keys, waterspouts like this are common across the waters of the Pacific Northwest. —Dr. Joseph Golden photo, courtesy of the National Oceanic and Atmospheric Administration/Department of Commerce

air, producing low clouds and fog. Typically cooler than the air above it, this so-called marine layer establishes a stable atmosphere that prevents intense surface heating necessary for thunderstorm development. So while plenty of moisture falls in the Pacific Northwest, the atmosphere often lacks the dramatic temperature contrast between the surface air and the air above it, preventing thunderstorms and tornadoes from forming.

It Rains All the Time: Seattle

"It rains all the time!" Everyone says this about Seattle. The oft-spoken words are somewhat accurate during winter: from late October through mid-May, western Washington is socked in with low clouds, fog, drizzle, and light rain. The winter rains make this the greenest place in the West, home to thick forests and flowering plants of all kinds. Not surprisingly, Olympia, Washington, ranks as the third cloudiest city in the United States, experiencing an average of 229 cloudy days per year, followed by Seattle with 227 cloudy days per year. It is not uncommon for days or even weeks to pass in the Puget Sound area without a sunny day, but it actually rains less here than in places you might expect to be drier.

Seattle's mean annual rainfall of 38.40 inches may seem like a lot for a western city, but surprisingly, both New York City, with more than 43 inches a year, and Miami, drenched by 55.10 inches annually, receive more rainfall. Seattle is soaked slowly throughout the winter with light rain and drizzle while the East Coast weathers heavier, short-duration rainstorms. New York has 120 days of rain while Seattle has nearly 160. So Seattle is not as wet as the long bouts of rain would make one think.

Summers are usually very dry in Seattle, but during fall and winter, moisture-laden mid-latitude storms blow into the region with regularity. The Aleutian low, a semipermanent low-pressure area over the Aleutian Islands of Alaska, intensifies and moves southward,

Record-Breaking Seattle Blizzards

When surface temperatures drop enough, Seattle has the potential to receive heavy snow as moisture-laden air from the Pacific travels over the land. Since moisture is rarely a limiting factor in this region, low temperatures cause heavy snow. Temperatures can drop when arctic air invades the region from British Columbia's interior.

On February 1, 1916, Seattle recorded its maximum snowfall ever in a twenty-four-hour period: 21.5 inches. Other parts of western Washington received between 2 and 4 feet of snow, and winds whipped snowdrifts as high as 5 feet. The storm crippled the region and halted all transportation. The Seattle office of the National Weather Service ranks this snowstorm as the ninth most significant weather event of the twentieth century for Washington. Similarly, a January 13, 1950, blizzard in the Puget Sound dropped 21.4 inches of snow on Seattle. The fierce blizzard claimed thirteen lives in the Puget Sound area alone.

reaching its maximum intensity midwinter. A semipermanent high-pressure area over the northern Pacific Ocean weakens and moves southward as well. The circulation around these two pressure centers creates a prevailing southwesterly and westerly flow of air directly into the Pacific Northwest. This flow of air, known as the polar jet stream, forms a fast-moving tube of air 30,000 to 40,000 feet above the ground. Storms develop along the jet stream over the central and western portions of the Pacific Ocean, where temperatures are warm enough to cause large amounts of seawater evaporation. These large mid-latitude storms sweep across the Pacific and hit the Pacific Northwest. As they meet the region, the moist air rises up the Cascades. It cools and condenses into rain on the west side of the Cascades. Often these storms are associated with a long fetch of atmospheric moisture streaming across the Pacific Ocean. So even after a main storm system has passed, a supply of moisture continues. This translates into more rain—sometimes lasting for days or even weeks—for the Puget Sound area.

The high latitude of Seattle—47.6 degrees north—makes winter days especially short. Imagine December 21, the shortest day of the year, as clear and sunny. Seattle would see only eight hours and fifteen minutes of sunshine—the sun rises at 7:55 AM and sets at 4:20 PM. However, if the day dawned cloudy, the overcast skies would block the sunshine and make the day seem even shorter.

Rainy days on end, coupled with limited daylight hours, affect people physically and psychologically. The winter blues, officially called seasonal affective disorder, or SAD, are especially severe in western Washington and Oregon. SAD symptoms range from sleepiness and heightened craving for starchy foods to depression and suicide. Since the best treatment for SAD is light, sunshine, and exercise, which helps the body produce vitamin D, experts advise people who live in regions with limited sunshine, to spend time outdoors, no matter what the weather is, and to keep their houses brightly lit.

Winter clouds and fog bring high humidity. Olympia's average annual relative humidity of 78

A powerful storm plows into the Pacific Northwest on October 27, 1999, bringing heavy rain, mountain snow, and high winds. A buoy off the California coast reported winds in excess of 85 miles per hour. Unhindered by topography, Pacific storms can form into classic, strong, mid-latitude cyclones like this. —Image courtesy of the National Oceanic and Atmospheric Administration

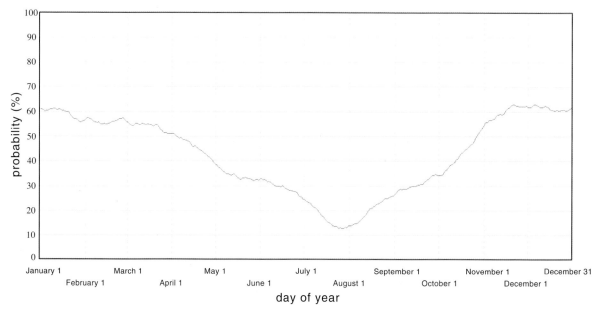

Probability of receiving 0.01 inch of precipitation in one day at Seattle, Washington. A 12 percent chance exists that this amount of precipitation will fall on July 28, while a 62 percent chance exists on November 20. —Courtesy of the National Oceanic and Atmospheric Administration, Western Regional Climate Center

Sacred Rain Rock

The Shasta Tribe of the Klamath Valley carved a 4,000-pound boulder from a flat-topped soapstone and used it during ceremonies. The boulder has forty-eight holes in its surface from times the Indians hammered on it to bring rain to the drought-stricken area. Considering the rock sacred—and effective—the tribe kept it buried to prevent deer from jumping on it and bringing unwanted rains.

At the end of one of the area's droughts, the tribe buried the rain rock in a deep hole in present-day Gottville, California, where it remained undisturbed for two hundred years. In the 1930s a road crew accidentally uncovered it. The rock now resides in the Fort Jones Museum in Fort Jones, California. Although the rain rock is not used today, people—non-Indians included—still request that the tribe cover the rock on occasion to prevent flooding rains.

percent makes it the second most humid city in the United States, behind first-place Quillayute, Washington—also the wettest U.S. city. Comfortably dry in the summer, Olympia's relative humidity in the winter soars so high that it significantly skews annual statistics. The relative humidity during an August afternoon averages a mere 50 percent, unlike the Southeast where summertime heat and humidity combine to create torrid conditions.

After a long wet winter nature rewards residents of the Pacific Northwest with a dry and warm summer. Undoubtedly, fewer people would live along the Pacific Northwest Coast if not for the delightful summers, dominated by warm days, cool nights, abundant sunshine, and prolonged dry periods. From July through September, the Puget Sound area sees an average of only fifty-four overcast days. Summer rains are rare as well—believe it or not, Salem, Oregon,

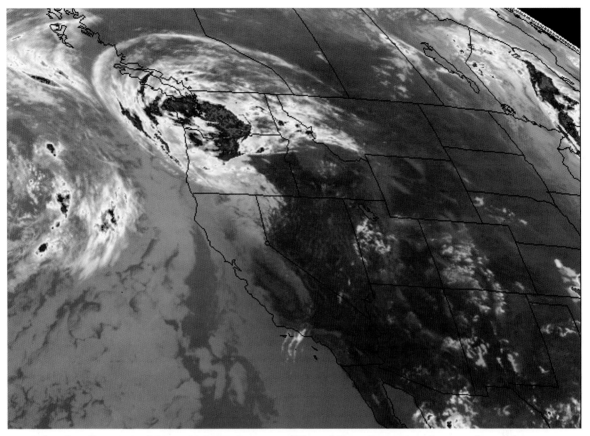

Infrared satellite image of August 20, 1997. Remnants of Tropical Storm Ignacio brought extremely rare, heavy summer rains to northern California, Oregon, and Washington. The storm occurred during a period with unusually warm offshore water associated with El Niño. The combination of warm water, an approaching low from the west, and a tropical storm set up an uncommon situation. The reds and purples indicate colder cloud tops and often heavier precipitation. The clouds arcing across southwestern Canada illustrate the storm's counterclockwise rotation.
—Image courtesy of the National Oceanic and Atmospheric Administration

222 miles south of Seattle and enjoying the same climate as the Puget Sound area, has never received measurable rain on July 12 since record keeping began back in 1892.

It takes an unusual weather situation to bring rain to the Puget Sound area during summer. For example, from August 18 through 20, 1997, the remnants of Tropical Storm Ignacio moved northward along the West Coast, producing large amounts of rare summer rain. The Puget Sound area received ½ inch of rain. Just north of Seattle 0.38 inches fell, establishing a new daily rainfall record for August 18.

The Blue Hole

The Olympic Mountains, in northwestern Washington, is the wettest place in the conterminous United States. Since the range parallels the coast and is perpendicular to the prevailing direction of incoming moist air, the western slopes receive phenomenal amounts of precipitation. Weather models estimate that areas of north-central Jefferson County receive more than 240 inches of precipitation a year. Most areas of the Olympic Mountains receive 144 to 168 inches of precipitation, including melted snowfall, annually. The high level of precipitation and humidity

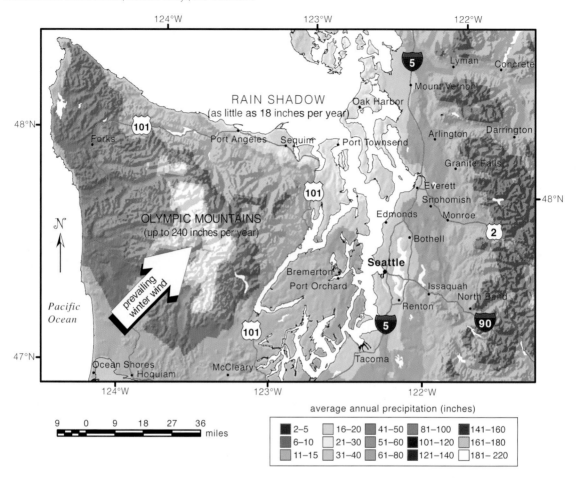

The Olympic Mountains, located on the northern section of the Olympic Peninsula, tower to nearly 8,000 feet and receive up to 240 inches of precipitation a year. The windward portions of the mountains receive the full force of storms moving inland from the Pacific Ocean, while on the leeward side of the Olympics the rain-shadow effect causes the average annual precipitation to be as low as 18 inches near Sequim.

on the western Olympic Peninsula supports the unusual and interesting vegetation of the Hoh Rain Forest. This temperate rain forest almost drips green—mosses and ferns drape immense Sitka spruce, western red cedar, bigleaf maple, and Douglas-fir trees. Moss mats on these trees may harbor entire communities of invertebrates and other organisms. Meanwhile, less than 50 miles away near Sequim, one of the world's most remarkable rain shadows receives so little precipitation that a species of cactus can grow there.

An area on the lee side of a mountain range that receives less precipitation than the area on the windward side forms a rain shadow. The southwestern slopes of the rugged Olympic Mountains host one of the largest temperate rain forests in the world. The mountains also shelter an area northeast of them from the unrelenting rains typical of the Pacific Northwest, creating an exceptionally dry climate for this region. Locals call the rain shadow the "Blue Hole," because from Sequim and the Dungeness Valley they can see clear, blue skies, while a shroud of fog and precipitation obscures nearby regions.

In the fall and winter, the polar jet stream sags southward over the Puget Sound area (see

A clear sky—the Blue Hole—shines over Sequim, Washington, and the Strait of Juan de Fuca downwind of the cloud-shrouded Olympic Mountains on June 1, 2004.

previous section, *It Rains All the Time: Seattle*). The southwesterly and westerly flow of the jet stream brings warm, moist air from over the ocean into the Pacific Northwest. The temperature of this air is often close to the that of the water, which is relatively warm compared to interior land temperatures at this latitude. Prevailing winds drive this moist air inland and up the slopes of the Olympic Mountains. As it rises from sea level to nearly 8,000 feet, the moisture condenses into clouds and precipitation, creating a wet season that begins in October, peaks in January, and gradually decreases by early summer.

The western Olympic Peninsula effectively wrings precipitation out of passing air masses, drenching the mountains with rain and leaving only a few sprinkles for the rain shadow area.

Areas around Sequim receive only 17 inches of precipitation a year. Eastern Clallam County is the driest area west of the Cascade Range. The rain shadow offers Washington residents a place to escape the rain without actually leaving the Pacific Northwest.

The Columbus Day Storm

Outside of the hurricane-stricken coastlines of the Southeast, the Pacific Northwest experiences some of the greatest windstorms in the United States. The strongest winds west of the Rocky Mountains whip coastal areas of northern California, Oregon, and Washington.

The main ingredient for these fierce westerly-to-southwesterly blasts is a strong pressure gradient that develops between an offshore storm with low pressure and higher pressure inland. The pressure difference between a deep—optimally less than 965 millibars—mid-latitude cyclone tracking northeastward off the West Coast and higher pressure southeast of the low causes a north-south-oriented pressure gradient to develop. The larger the pressure difference, the stronger the winds. For instance, a pressure difference of 9 to 12 millibars between North Bend on the central Oregon coast and Quillayute along the northern Washington coast can generate 20- to 40-mile-per-hour winds. An 18 to 24 millibar difference can generate 50- to 70-mile-per-hour winds, which are typical of moderate to strong windstorms. Major windstorms, such as the Columbus Day Storm, are caused by 25- to 30-millibar gradients and blow well over 120 miles per hour.

At inland locations, such as the Willamette Valley and Puget Sound, a weaker gradient can create strong windstorms since other factors, including low-level jet streams, wind channeling, and convection, contribute to the strong winds. A pressure gradient of 9 to 10 millibars occurring between Portland and Eugene—called the "sweet spot" by local meteorologists—consistently produces strong winds in the Willamette Valley.

In Washington, low-pressure areas can generate a powerful lee eddy—a compact cyclone that occurs when terrain squeezes the atmosphere into a smaller area—east of the Olympic Mountains. With relatively higher pressure in Oregon, these cyclones and the induced pressure gradient can whip up dangerously strong southerly winds of 75 to 90 miles per hour across the Seattle-Portland corridor. Severe events, with wind gusts of 80-plus miles per hour, hit the area every fifty years, and weaker winds of 40 to 70 miles per hour blow each winter.

Coastal regions of Washington and especially Oregon receive the most brutal battering—60-mile-per-hour winds lash the area each winter. Along the Oregon Coast, winds howl at 50 miles per hour or more an average of 250 hours a year.

Puget Sound Convergence Zone

Sometimes Seattle's morning traffic and weather report will warn travelers about "areas of heavy rain or snow," or "mixed rain and snow in the convergence zone." A traveler passing through might confuse the "convergence zone" with some notorious area of tangled traffic. In reality, air masses converge just north of Seattle at a site meteorologists call the Puget Sound Convergence Zone, or PSCZ. This unique weather phenomenon occurs several dozen times a year in the winter, spring, and early summer, and it is known well by locals and clearly understood by the team of experts at the University of Washington who first discovered it in 1980.

A PSCZ event occurs when the atmosphere is unstable and low-level westerly wind is split into two distinct currents by the Olympic Mountains: one is channeled to the northeast through the Strait of Juan de Fuca and the other to the southeast through the Chehalis Gap west of Olympia. If the low-level wind is shallow and mild, the winds split instead of traveling over the Olympics. (Although the moist wind channeled through the Strait of Juan de Fuca travels across the Olympic Mountain rain shadow region, no mechanism—such as a mountain range or cold front—forces the water vapor to cool, condense, and turn into precipitation; therefore, even under PSCZ conditions, the rain shadow region avoids any significant precipitation.

As the winds curl around the domelike Olympic Mountains, they become two channels of westerly blowing wind. As the two channels of wind encounter the north-south-oriented Cascades, they are deflected. The Strait of Juan de Fuca wind becomes a northerly wind

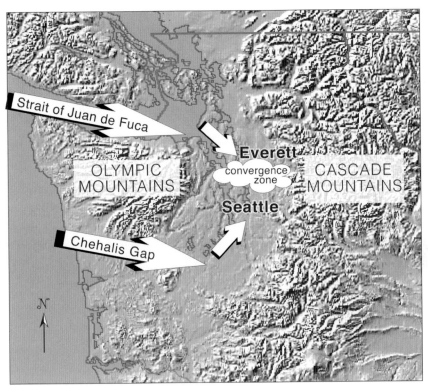

A simplified illustration of the Puget Sound Convergence Zone

while the Chehalis Gap wind becomes a southerly wind. These moisture-laden winds collide suddenly and violently—typically between northern Seattle and Everett—and erupt into an east-west elongated zone of heavy rain and/or snow, gusty erratic winds, cold temperatures, and even lightning and thunder.

Stormy episodes typically last two to four hours, but an entire PSCZ event can last two or three days. The wind's boundary meanders to the north and south over this period. The most intense PSCZ events occur during spring and early summer, when low-level winds are not too strong and moisture is still plentiful.

The PSCZ can generate dramatic weather variations in a very short distance. Its storms cause accidents, halt traffic, and commonly bury affected areas with up to 8 inches of wind-driven snow, while surrounding areas might only experience light rain. A notorious PSCZ episode occurred on December 18, 1990. A classic convergence zone developed in north Seattle, dumping 6 to 14 inches of snow, while the Everett and Sea-Tac Airports reported only 2 to 4 inches.

Other areas of the West, including Denver and Las Vegas, have similar topographically induced convergence zones, but the PSCZ ranks among the most famous.

Winds of 80 or 100 miles per hour commonly blow. Cape Blanco, Oregon, notorious for its strong winds, is commonly the windiest location on the West Coast, with or without a storm. But in 1962, a vigorous storm brought winds of nearly 200 miles per hour.

The Columbus Day Storm, also called the Big Blow, hit western Washington and Oregon on October 12, 1962—arguably the mightiest such storm to ever hit this area. You might think of it as the Pacific's "Perfect Storm." This unique storm began eight days earlier as a powerful typhoon named Freda, near Wake Island in the western Pacific Ocean. By the time the storm traversed the Pacific, it had lost its tropical-storm characteristics—energy from warm sea surface temperatures, a discernible "eye" in the center of the low, and relatively warm temperatures aloft—but not its tremendous power. At one point, the storm registered a central pressure of approximately 960 millibars (28.35 inches), among the lowest pressures ever to visit the Pacific Northwest Coast. For comparison, a December 12, 1995, storm pushed Astoria's air pressure to an all-time state low of 966 millibars (28.53 inches), comparable to the central pressure of a Category 2 hurricane (see the Saffir-Simpson scale in *Hurricane Alley of the West* in *chapter 6*).

A Fitting Name

The famous British navigator Captain James Cook discovered and named Cape Foulweather, near Depoe Bay, Oregon, in 1778. The cliffs near Cape Foulweather, perched 500 feet above sea level, are some of the highest points along the Oregon Coast. Captain Cook first sighted the mainland of North America there, and one of the cape's characteristically sudden storms struck upon his arrival, almost ending the historic expedition. The name he gave this rugged landmark reflects the fierceness of the storm.

Peak Wind Gusts in Oregon and Washington during the Columbus Day Storm

LOCATION	WIND SPEED (miles per hour)
Cape Blanco, Oregon	195 (estimated)
Mount Hebo Air Force Radar Base, Oregon	170
Newport, Oregon	138
Corvallis Municipal Airport	127
Downtown Portland (Morrison Street Bridge)	116
Troutdale Airport	106
Portland International Airport	104
Renton, Washington	100
Bellingham, Washington	98
Astoria, Oregon	96
Downtown Portland (KGW radio station)	93
Hillsboro, Oregon	90
Salem, Oregon	90
Whidbey Island, Washington	90
Eugene, Oregon	86
North Bend, Oregon	81
Klamath Falls, Oregon	65
Sea-Tac International Airport	65
Roseburg, Oregon	62
Lakeview, Oregon	58
Medford, Oregon	58
Redmond, Oregon	47
The Dalles, Oregon	29

During the Columbus Day Storm, Portland International Airport reported its all-time-highest wind gust of 104 miles per hour, and a 116-mile-per-hour gust rattled the downtown Morrison Street Bridge. Officials measured a gust of 65 miles per hour at Sea-Tac International Airport. A 170-mile-per-hour gust was clocked on Mount Hebo, just southeast of Tillamook,

Strong winds of the October 12, 1962, Columbus Day Storm knocked this tree over. —Photo courtesy of the City of Portland Archives and Records

Oregon, before the wind gauge blew away. This represents the highest unofficial wind gust ever recorded in the state of Oregon, even though weather observers estimated that 195-mile-per-hour winds blew at Cape Blanco at the height of the storm. Although these winds were estimated, current wind equipment at Cape Blanco suggests it experiences the strongest winds of any place along the West Coast.

Wind-whipped foam turned the ocean white, while huge waves crashed into the coast, threatening to wash away communities. The Columbus Day Storm caused more destruction in the Pacific Northwest than any other windstorm in recorded history. The winds jerked out one-hundred-year-old trees, ripped down power lines, broke windows, destroyed crops, devastated area homes and buildings, and even sank a few pontoon bridges. Some western Oregon cities were without power for three weeks. The storm claimed thirty-eight lives, inflicted $170 to 200 million in

Willamette Valley Gusts

Ordinarily, for winds to reach 50 miles per hour in the Willamette Valley and 80 miles per hour along the coast, the atmospheric pressure at Astoria, Oregon, must drop to about 985 millibars (29.09 inches). This is usually low enough to cause a significant pressure gradient to develop between low pressure offshore and higher pressure over southern Oregon.

damage ($1 to 1.2 billion in 2002 dollars), and knocked down over 11 billion board feet of trees, approximately two years of cut for Oregon. The Portland office of the National Weather Service ranks the storm as Oregon's third most significant weather event of the twentieth century. Likewise, the National Weather Service office in Seattle ranks the windstorm as Washington's number one weather event of the century.

A Stormy Getaway for Two

Rushing to the coast to watch winter storms has long entertained Oregonians. Known as "storm watching," it has now caught on with tourists who show up on Oregon's beaches months after summer vacationers have abandoned the area. Coastal hotels even offer winter storm-watching packages with names such as "Romance of the Storm" and "Stormy Weather Getaway," which include rooms with fireplaces and seaside balconies. The winter storms' long fetch of winds, which blow at 50 to 150 miles per hour, generate huge waves that are the main attraction for storm watchers. Along with massive waves, the storms can bring extremely strong winds, flooding rains, and even lightning and cold air funnels for a storm viewer's enjoyment.

Thermal Low: Willamette Valley

A summertime weather phenomenon known as a *thermal low* brings unseasonably warm temperatures to Oregon's Willamette Valley. Unlike ordinary lows, which bring rain or snow, a thermal low, as its name infers, is a shallow area of intense heat with little or no precipitation.

A thermal low is caused by hot, rising surface-heated air, which induces a low-pressure circulation. The counterclockwise circulation around the low is strongest nearest the surface, where the column of rising air rises fastest, and weakens with altitude. A thermal low typically extends no more than a mile or two above the ground and can cover an area from central Oregon to central California.

Thermal lows, also called *heat lows, thermal troughs,* or simply *trofs,* develop across sun-baked deserts of southern Arizona, southern Nevada, and southeastern California. They get their name because in three-dimensional space, low pressure causes troughs to form in the atmosphere and air tends to blow toward them—just as water flows toward a low spot. Though these troughs form over the southwestern deserts, strong high-pressure systems moving south out of Canada into the western Great Basin actually push or advect the thermal trough northward over the central valleys of California and into Oregon. The low's counterclockwise winds rotate hot air from the desert into the Willamette Valley. These northeasterly breezes warm and dry through compressional heating as they descend the Cascades, further heating—sometimes to over 100°F—the already hot air. In some cases, a trough of low pressure will extend as far north as Seattle.

The intense heat associated with thermal lows warms western Oregon one to four times a year. During these events, the hottest day occurs when the thermal low's axis, or trough axis, is directly overhead, where the hottest air of the thermal low exists. Not surprisingly, most of the all-time record high temperatures across the Willamette Valley occurred during thermal lows. In general, a thermal low can drive surface temperatures into the 90s and sometimes the low 100s in the Willamette Valley, which is hot considering the average summertime temperature is about 80°F. When the trough axis moves along the coastline, even the coast can see temperatures in the 90s; a normal summertime temperature reaches about 65°F. On August 13, 2002, Newport, Oregon, endured its all-time high temperature of 99°F. As the trough axis moved inland, the temperature at Oregon City reached 104°F! One of the longest thermal-low heat waves in the Willamette Valley happened July 19–23, 1994, when Portland

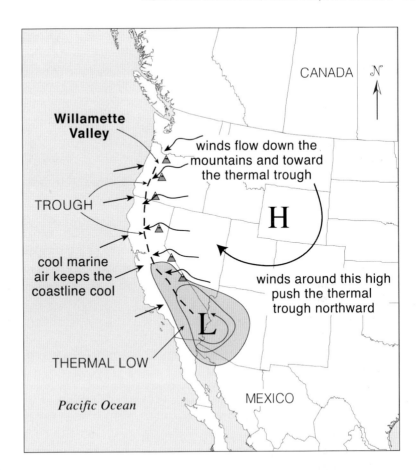

Simplified illustration of a thermal low, its associated trough of low pressure (dashed line), and the winds associated with a Rocky Mountain high-pressure area that pushes the trough northward

suffered through five consecutive days of hot temperatures: 102°F, 103°F, 101°F, 97°F, and 96°F, respectively. At the same time the coast enjoyed pleasant temperatures near 70°F.

Marine Push

Also known as "marine surges" or "natural air conditioning," marine pushes are summertime occurrences that bring cool, marine air and fog to the western valleys of Washington and Oregon. The marine push cycle begins during the day when a high-pressure system residing over the Pacific off the coasts of Oregon and Washington lowers pressure inland. Most of the inland low pressure is associated with heat rising off Washington and Oregon as surfaces warm during the day. Sometimes, however, the low pressure is associated with thermal lows. Fog and low clouds develop along the coast as the warm inland air condenses over the much cooler Pacific Ocean. The rising hot air creates a vacuum inland, and at night the high-pressure system pushes cool marine air through the gaps of the Coast Ranges of Oregon and the Olympics in Washington to fill the void left by the rising air. Usually the marine push will bring morning fog pushed in from along the coast, low clouds, and cooler temperatures to Washington's and Oregon's western valleys, but by afternoon skies will clear and the process will start over again.

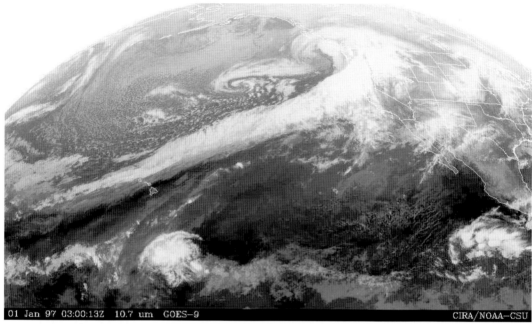

01 Jan 97 03:00:13Z   10.7 um   GOES-9   CIRA/NOAA-CSU

January 1, 1997, infrared satellite image showing a massive, flood-producing Pineapple Express stretching from west of Hawai'i to the West Coast. The torrential rains and warm temperatures associated with this and other expresses in December 1996 and January 1997 caused $3.4 billion in damage and thirty-six deaths across California, Washington, Oregon, Idaho, Nevada, and Montana. —Image courtesy of the National Oceanic and Atmospheric Administration

A thermal low typically lingers in western Oregon only a day or two. Eventually, the heat low weakens and cooler temperatures return to western Oregon. Usually the upper-level wind pattern shifts enough to decouple and move the trough of hot air east. The upper-level winds change when a weak low-pressure system approaches the coast or the high-pressure system over the Great Basin shifts east, allowing cooler air to enter the region. Sometimes a persistent series of marine surges can cause the thermal low to weaken as the cool air mixes with hot air, but it can take several days for the marine air to fully invade the valleys of western Oregon and Washington. Regardless, as the trough pushes east of the Cascades, the winds change from easterly to westerly, bringing a rush of cooler ocean air and sometimes triggering a marine push that brings an abrupt end to the heat spell.

The Pineapple Express and the Flood of '96

During the West Coast's wet season, October through May, parts of the Pacific Northwest and California are drenched by the Pineapple Express, also known as the Pineapple Connection. This waterlogged weather pattern develops when upper-level winds blow from the southwest and pump warm, saturated air from near the Hawaiian Islands northeastward into California, western Oregon, and western Washington. During these events the polar and subtropical jet streams propel this strong flow of moisture-laden clouds into the region. One good way to think of the Pineapple Express is to imagine a gigantic fire hose wildly and relentlessly spraying the Pacific Coast.

A Pineapple Express develops when the Aleutian low strengthens and shifts southward and the

Aerial view of homes flooded in the subdivision of Arboga, south of Marysville and north of Sacramento, California, on January 5, 1997. Heavy rains and massive snowmelt caused a levee along the Feather River to break on January 2, 1997, flooding this neighborhood and many others. —Photo courtesy of Robert A. Eplett, California Governor's Office of Emergency Services

Floodwaters surround a home along Interstates 5 and 205, west of Stockton, California, after heavy rains from a Pineapple Express broke a levee. Owners purposely built this home on a mound to raise it above the floodplain. —Photo courtesy of Robert A. Eplett, California Governor's Office of Emergency Services

Flood damage to a California road, January 1, 1997 —Photo courtesy of the U.S. Army Corps of Engineers, Sacramento District

polar jet stream, rounding the southern and eastern side of the low, meets the subtropical jet stream. This concentrated, large-scale flow across the ocean advects a significant amount of subtropical moisture in a southwest to northeast orientation. The temperature contrast between the cold Aleutian low and the warmer air south of the subtropical jet stream causes a cold front to form along the combined polar and subtropical jet stream. The cold front becomes the focus of the moisture stream, and the heaviest precipitation falls just south of the cold front. However, sometimes the moisture will overrun the colder air and drop as heavy snow just north of the cold front.

The express of clouds, warm air, and rain spread extremely wet conditions across much of

the West Coast with the area between Portland, Oregon, and San Francisco, California, typically receiving the most direct hit. The resulting precipitation and warm temperatures can profoundly affect road travel, making bridges and roads impassable because of floodwaters and mudslides.

The West Coast typically experiences one to four Pineapple Expresses a year. It can rain and snow at higher elevations for three to five days during a Pineapple Express, but the warm air of these events can melt high-elevation snow and aggravate any flooding its rain has already caused. Often a rain-on-snow event—rain falling on top of snow—occurs, which severely floods streams and rivers with massive amounts of water. The most devastating Pineapple Expresses remain stationary for more than two days, sustained by a long fetch of moisture-laden winds blowing across the Pacific Ocean. Normally, atmospheric winds cannot sustain a long fetch of moisture directed at the same location for more than a day.

The frequency of Pineapple Expresses during the winter of 1996–97 brought the worst flooding to Northern California in decades. Mudslides and washouts closed some roads for months. Residents dubbed it the "Year of the Perpetual Car Wash." Numerous rainfall records were smashed.

Between December 26, 1996, and January 3, 1997, a Pineapple Express dumped huge amounts of moisture on the entire West Coast. California locations received tremendous amounts of rain: Napa received 10.82 inches and Sacramento 5.67 inches. Its aftermath became known as the New Year's Flood. The express, coupled with tremendous snowmelt, sent rivers and lakes over their banks all along the West Coast and in the Sierra Nevada, particularly in the Sacramento Valley where rivers drained out of the Sierras. The floods breached levees across California, and saturated hillsides released countless mudslides. Floodwater submerged 300 square miles of California, including the Yosemite Valley, which flooded for the first time since the winter of

1861–62. For weeks after the rains had stopped, rivers continued to flow over their banks, and flood damage and mudslides left major roads impassable. Mudslides and floodwater cut the trans-Sierra U.S. 50 in five places. A system of levees spared Sacramento from the worst of the flooding, but many nearby towns suffered. Fifteen California rivers experienced significant flooding, and three rivers—the Cosumnes, South Fork of the American, and Napa—approached or exceeded historical peak flows.

Parts of the central Sierra Nevada received more than 40 inches of rain—yes, rain, not snow—in a week. Blue Canyon, California, just west of the crest of the Sierra Nevada along Interstate 80, received 39.34 inches. Lake Tahoe rose to 6,229.4 feet, its highest level since 1917. According to the U.S. Geological Survey, the average surface of the lake is 6,225 feet above sea level.

Emergency-management officials reported that the New Year's Flood forced as many as one-half million West Coast residents—Oregon and Washington included—to evacuate at some point during its course. On January 20, 1997, officials declared forty-three of California's fifty-eight counties disaster areas. The National Weather Service ranks the flood as California's fifth and Nevada's second most significant weather event of the twentieth century. By the time floodwaters receded and damages were tallied, the flood had caused thirty-six deaths, countless injuries, and an estimated $3.4 billion ($3.8 billion in 2002 dollars) in damage, making it one of the few billion-dollar weather events in the West.

The Brookings Effect: Oregon

Afternoon temperatures in and near the Oregon coastal community of Brookings rise into the 80s and 90s, while temperatures along the remainder of the Pacific Northwest coast hover in the 60s. The Brookings Effect, which can occur at any time of year, causes this unique, localized,

Oregon under Water

Another major Pineapple Express episode occurred between February 5 and 9, 1996, and the "hose" was directed at the northern half of Oregon. After a week of cold, snowy weather, up to 30 inches of rain fell in five days in Oregon's Coast Ranges and across southwestern Washington. Melting snow and heavy rains flooded rivers, and floodwaters running on frozen soil quickly turned into mudslides that ran across roads. This Express unleashed the worst flooding in thirty years on parts of northwestern Oregon. In all, the storm killed six people, forced more than thirty thousand people to evacuate their homes, and caused millions of dollars in damage. Areas hit the hardest included Tillamook, Keizer, and Tualatin, Oregon, and Woodland, Washington. Westerners remember this event as the Great Flood of '96.

In the same year, at least forty-one locations in Oregon measured their highest annual precipitation ever because of Pineapple Expresses. Laurel Mountain, at an elevation of 3,590 feet and 22 miles west of Salem in the Coast Ranges, received the most precipitation ever at an Oregon location: 204.12 inches. This shattered the previous state record of 168.88 inches set in 1937 at Valsetz, just 10 miles to the southwest, and established a new rainfall record for the contiguous United States. This new record surpassed the 1931 record of 184.56 inches set at Wynoochee Oxbow, Washington. Even more amazing about the Laurel Mountain record, more than 90 inches came in just two months: 42.30 inches in February and 49.57 inches in December. These staggering monthly amounts still fall short of the U.S. monthly rainfall record of 71.54 inches, which fell at Helen Mine, California, in January 1909. Although notoriously wet, in 1996 the Pacific Northwest proved too wet even for the natives.

Extreme Weather Observer

In 1890, under the Organic Act, the National Weather Service (NWS) Cooperative Observer Program (COOP) began as a climate-observing network of, by, and for the people. Today, over eleven thousand volunteers record daily weather observations around the nation. The COOP mission, as outlined in the Organic Act, pledges:

> To provide observational meteorological data, usually consisting of daily maximum and minimum temperatures, snowfall, and twenty-four-hour precipitation totals, required to define the climate of the United States and to help measure long-term climate changes.
>
> To provide observational meteorological data in near real time to support forecast, warning, and other public service programs of the NWS.

To recognize and reward excellence in weather observation, the National Weather Service developed the Cooperative Weather Observer Awards Program. Among the awards is the Earl Stewart Award, an award named for a cooperative observer at Cottage Grove, Oregon. Stewart recorded the weather for seventy-eight years, from 1917 until 1996. A meticulous observer, he rarely missed an observation, trusting only one friend to take observations in his absence. In 1996, Earl died, ending one of the longest weather-observing streaks in U.S. history. The award, which only he has received to date, recognizes those who serve as observers for seventy-five years or more.

Earl Stewart and his wife Dorothy at their home in Cottage Grove, Oregon, where Earl recorded the daily weather for seventy-eight years, from 1917 until 1996
—George Taylor photo

and short-lived temperature event. Locals and meteorologists observed and wondered about the now-termed Brookings Effect for years, until a distinguished TV weathercaster in Portland, Jack Capell, eventually figured out that the terrain and ocean waters around Brookings work together to cause this weather oddity.

Brookings sits in a large cove surrounded by mountains on three sides and the open ocean on the fourth. The prevailing surface and low-level winds from the north-northwest sometimes cause an eddy to develop in the lee of the headlands just northwest of Brookings. When the surface pressure falls drastically at Brookings, it indicates the onset of the miniature eddy. Since the lower pressure occupies less vertical space than the surrounding air, it draws in warmer air (warmed by adiabatic warming) from the higher inland elevations.

The Coast Ranges east of Brookings rises to 5,000 feet, with Mt. Emily at 2,926 feet just 8 miles west-northwest of town, so the air drops as much as 5,000 feet as it descends these mountains. The air is funneled down through the Chetco River valley and into Brookings, arriving as a warm easterly breeze. As the air drops it encounters the higher air pressure of lower elevation and compresses and warms. This phenomenon, known as *adiabatic warming,* warms air 5.5°F for every 1,000 feet of descent. Theoretically, air at 70°F in the Coast Ranges could be 92 to 97°F by the time it reaches the coast. The descending warm air heats the Brookings area, and eventually the low warms enough to be classified as a heat low. The heat low heats even more when lower pressure from California noses northward, creating an overall offshore flow across the southern Oregon

coast, and sometimes providing a welcome heat wave as far south as Point St. George, California, and north to Gold Beach, Oregon.

Sometimes the Brookings Effect can raise temperatures in a matter of minutes. For example, on December 20, 1999, Brookings had a temperature in the 60s, with relative humidity at 65 percent. A northeasterly wind picked up and in fifteen minutes the temperature rose to 78°F and the humidity dropped to 15 percent!

The Brookings Effect, a land- and sea-assisted microclimate, can grossly misrepresent the weather conditions along the southern Oregon coast. The weather anomaly is so dramatic that the National Climatic Data Center's sophisticated quality-control software often flags Brookings's daily temperatures as suspect. Comparisons with all other nearby weather stations lead the software

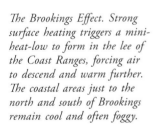

The Brookings Effect. Strong surface heating triggers a mini-heat-low to form in the lee of the Coast Ranges, forcing air to descend and warm further. The coastal areas just to the north and south of Brookings remain cool and often foggy.

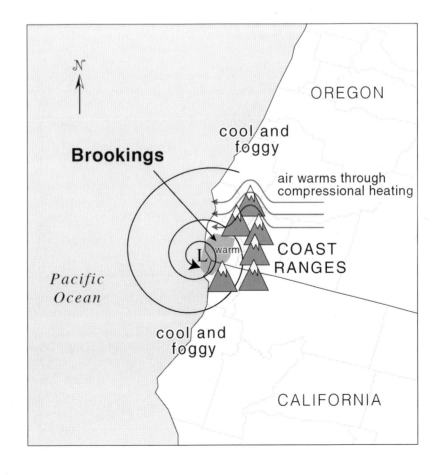

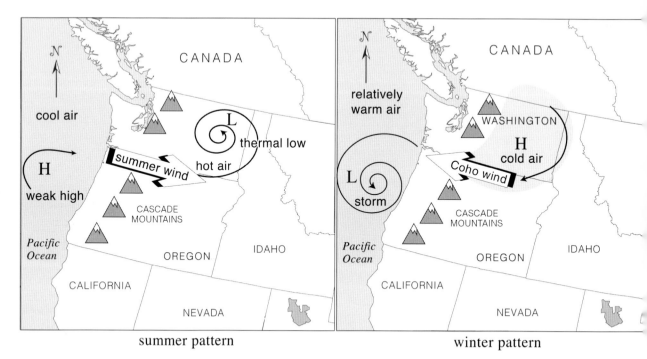

summer pattern   winter pattern

Simplified illustration of the push and pull of air through the narrow Columbia River Gorge. In summer, a small but powerful thermal low forms over eastern Washington. Meanwhile, the air over the cool Pacific induces a relatively weak high-pressure area, causing a stiff westerly wind. In winter, the opposite occurs: A low forms, or moves, just off the Oregon Coast while a cold air–induced high-pressure area develops over eastern Washington causing an easterly wind.

to believe the Brookings temperatures register erroneously. And while the Brookings Effect is among the most reliable small-scale weather effects in the country, the town also gets its share of cloudy, foggy days—even when the remainder of the northern Oregon Coast enjoys warmth and sun. You could call it the "reverse Brookings Effect" when the surrounding mountains actually trap fog and low clouds being advected into the area by southwesterly winds. However, fog and clouds are rare at Brookings—due to the Brookings Effect—compared to other coastal locations, giving rise to the area's nickname as the Banana Belt of the southern Oregon coast.

Coho Winds and Silver Thaw: Columbia River Gorge

Known for its spectacular beauty, the Columbia River Gorge's magnificent stone walls funnel persistent, sometimes damaging, and often cold easterly winter winds. "Gorge winds," as some call them, are channeled through this gap between Washington and Oregon, concentrating into fierce onslaughts that blast the Portland-Vancouver area with chilly 40- to 80-mile-per-hour winds several times each winter.

Until recently, these pesky winds lacked an official name. But in November 1997, a Portland television station and The Oregonian teamed with the Oregon chapter of the American Meteorological Society to launch the "Name Our East Wind" contest. Some 7,000 entries were submitted with a total of 2,424 different names. In the end, judges chose the name "Coho." Wild and fast swimmers, Coho salmon are much like the gorge winds. The name is indigenous to the Pacific Northwest and easy to pronounce and spell. Coho also relates to the widely known westerly Chinook winds of the Rockies, also named for a type of salmon.

Since pressure gradients are generally larger in winter than summer, winter storms are usually stronger than those in summer. Strong winds in the gorge correlate with the strength of the pressure gradient through the gorge. A strong pressure gradient can be caused by a strong low offshore, a strong high inland, or a combination of both. In the Columbia River Gorge, a strong pressure gradient means fast wind speeds. Since winds generally blow from areas of high pressure to low pressure, cold air in a high-pressure system over eastern Washington is drawn westward through the gorge by Pacific storms approaching the coast. The low pressure associated with approaching storms causes a vacuum of air, so inland air is drawn into the low-pressure area in an attempt to fill the void. Furthermore, counterclockwise winds around the low only increase the flow of air into the low. Once set in motion, the winds are squeezed through the gorge and can reach speeds of 80 miles per hour, causing structural damage and even blowing trucks off Interstate 84.

Imagine eastern Washington as a bowl of cold air, with the cities of Pasco, Richland, and Kennewick in southern Washington as a hole in the bottom that empties icy winds into the gorge. The cold, dense air rushes through the gorge, warming only slightly along its journey to Portland. It spills into the Portland area, chilling the air and turning any liquid precipitation into snow and ice, which is known locally as "Silver Thaw."

Katabatic Winds around the World

Fierce winter katabatic winds are observed throughout the world, including in Greenland, Antarctica, Yosemite National Park, and the northern Adriatic Coast in eastern Europe. In fact, Cape Denison on the Antarctic Coast has the world's highest average wind speed—50 miles per hour—due largely to katabatic winds.

Winds in the katabatic family (*katabatic* means "to fall") resemble the cold bora winds that blow in Yugoslavia; a significant descent is necessary for such winds. A Coho wind descends very little as it travels through the gorge, yet it still has katabatic wind characteristics: it originates from higher terrain and is cold. More broadly, the Coho is known as a *gap wind*. Although air warms by compression as it blows downhill, Coho winds have much lower temperatures than the mild marine air they replace on the Pacific Coast since they descend little and experience limited adiabatic warming. Because Coho winds originate in eastern Washington where the temperature dives below zero, they only achieve 20 to 30°F temperatures by the time they hit the coast. Factor in windchill, and the actual, experienced temperature in Portland can sometimes drop to −20°F, hence the wind's other nickname, the "Portland Refrigerator."

Cold temperatures and fast winds aren't the only thing Oregon and Washington residents dread with the onset of Coho winds. On average, more severe ice storms glaze Portland, Oregon, and the Columbia River Gorge than any other location in the West. The ice storms, or Silver Thaws, wreak havoc in the gorge eight to twelve times a year but only about two to four times a year in the greater Portland area. Dreaded more than any other type of winter weather, ice storms can cause major damage and bring daily life to a standstill.

A Portland-style ice storm begins when easterly Coho winds blow through the gorge and into approaching Pacific storms. The abundant moisture streaming in from the Pacific Ocean meets the cold air coming from the east, creating freezing rain. The freezing rain can encase every east-facing object with 1 to 4 inches of ice. Ice can literally freeze parked cars and trucks to the road, and it can encrust an automobile antenna, making it grow to the diameter of a baseball bat. Gorge ice storms can quickly transform Interstate 84 into an ice-skating rink. The powerful winds, blowing at 30 to 70 miles per hour, can easily rip down ice-laden power lines and trees.

In winter, upper-level winds can aim a seemingly endless number of potent Pacific storms at this region, prolonging ice storms for up to five days. On average, Portland International Airport experiences 13.4 hours of freezing rain per year, but storms have lasted more than 40 hours. Between January 18 and 20, 1950, freezing rain fell for 41 consecutive hours!

As the cold interior air moves farther away from the gorge, it encounters increasingly warmer marine air. Much of downtown Portland and the remainder of the western valleys usually escape the brunt of Silver Thaws since the marine air mixes with the cooler air, keeping temperatures above

Gap Winds

Winds such as the Coho, whose velocity and path are dictated by a gap in a mountain range, are sometimes referred to as *gap winds*. While the Columbia River Gorge is the largest gap through the Cascades and the area most prone to severe boras in the Northwest, boras can gain devastating speeds in smaller gaps as well.

For example, the Enumclaw wind frequently rattles the foothill and canyon communities of western Washington— particularly Enumclaw—with 50- to 100-mile-per-hour winds each winter. On Christmas Eve 1983, this wind rushed through Stampede Gap, reaching 100 miles per hour near Enumclaw. The most severe windstorm of its kind ever experienced in the Pacific Northwest, it flattened several homes and inflicted $27 million ($48.7 million in 2002 dollars) in damage to northern Oregon and western Washington.

Similar to the Enumclaw wind, the Fraser Canyon wind blows frigid air from the interior of Canada through Fraser Canyon into northern Washington. The combination of this arctic air and warm, moist air associated with Pacific storms can result in severe snowstorms in and around Bellingham, Washington. If the cold air is relatively shallow in central British Columbia, it will have the ability to pass through only the low-elevation gaps such as Fraser Canyon. However, if the cold air is deeper, it will pass through higher gaps, such as the Enumclaw Valley, and rush into the lowlands of the coast.

One such snowstorm, caused by a deep layer of cold air, hit Bellingham and much of northwest Washington on December 29, 1996. Powerful northerly winds of 70 to 90 miles per hour whipped a fresh 2-foot snowfall into 12-foot drifts in Bellingham. The temperature during the morning hovered in the low teens, while the windchill dropped it to −30 to −50°F. Not a single road in Whatcom County, Washington, was passable for a time—even the snowplows got stuck. The blizzard paralyzed northwestern Washington for days.

A December 1996 ice storm coats everything in several layers of ice near Portland, Oregon.
—Tyree Wilde photo, National Weather Service, Portland, Oregon

freezing. Freezing rain usually occurs in the vicinity of the gorge, including areas around Corbett, Troutdale, and often as far west as Portland International Airport; however, strong Coho winds can cool the entire Willamette Valley of western Oregon to temperatures at or below freezing, resulting in extremely large, damaging ice storms.

The easterly Coho winds not only are stronger but also are more frequent than the westerly summertime wind in the gorge ("Coho" describes only the cold, easterly, winter wind and not its summertime counterpart). In summer the heated land surface over eastern Washington and Oregon induces a low-pressure area of warm, rising air known as a heat low (see *Thermal Low* earlier in this chapter), while higher pressure or sinking air resides over the cold ocean waters of the Pacific Ocean. The wind blows from high pressure off the coast, through the gorge, and into the inland low-pressure area. The winds in the summer typically blow at 10 to 20 miles per hour but blow more consistently than the Coho winds since the pressure gradient is less intense. However, if a strong heat low develops, winds can kick up to 50 miles per hour and create areas of blowing dust. Strong summer winds following a cold front or strong marine pushes can enhance heat lows; the strong westerly winds associated with these events cause winds to blow down the east side of the Cascades. The winds warm through compressional heating, strengthening the low over eastern Washington and increasing the pressure gradient and the speed of the gorge wind.

The westerly wind blows at an annual average of 13 miles per hour at Hood River, Oregon. This approaches speeds of the notoriously windy Great Plains of Texas and Kansas, where the average annual wind speed reaches 15 miles per hour. The combination of the wide Columbia River and persistent summer winds makes Hood River one of the premier windsurfing places in the world. On the other hand, it can be one of the most treacherous places to drive during a winter storm.

Pileup on Interstate 84

A severe dust storm driven by westerly gorge winds caused a forty-eight-car pileup on Interstate 84 near Boardman, Oregon, killing six people on September 26, 1999. The strong wind and blowing dust reduced visibility to zero at times.

Notorious Ice Storms in the Portland Area

November 18–22, 1921: A lengthy ice, snow, and sleet storm clogged the Columbia River Gorge. The storm buried The Dalles, Oregon, under an icy mass 54 inches deep that totaled 8.90 inches of liquid precipitation. The average annual precipitation at The Dalles is only 13.92 inches.

January 18, 1950: An all-day sleet-, rain-, and snowstorm turned to freezing rain, leaving hundreds of motorists stranded in the Columbia River Gorge. Trains rescued them, but even rail traffic experienced considerable difficulty and many delays getting through the gorge. Freezing rain downed trees and power lines, creating widespread power outages across northwestern Oregon. The storm caused hundreds of thousands of dollars in damage. The National Weather Service declared the weather of January 1950 as the eighth most significant weather event of the twentieth century in Oregon and the third most significant in Washington.

January 17–19, 1970: One of the most damaging ice storms in recent years struck the Columbia River Gorge, causing millions of dollars in damage—$6 million ($27.7 million in 2002 dollars) in damage

to the Hood River Valley area alone. Up to 1 1/2 inches of clear, solid ice coated every exposed surface in the region.

January 9–10, 1979: A bad ice storm covered Portland and the western Columbia River Gorge in a heavy glaze of clear ice. Seventy-five thousand homes and businesses lost electricity, and Interstate 84 remained closed for days.

December 24–30, 1996: A series of fierce winter storms walloped the Columbia River Gorge, plastering 1 to 4 inches of ice onto every east-facing object. Trucks froze to Interstate 84 and stayed stuck for five days. Easterly winds of 20 to 40 miles per hour snapped ice-coated power lines, knocking out power to 122,000 homes. Portland International Airport measured nearly 4 inches of precipitation during this period, and the storm left one person dead.

January 6–7, 2003: A major ice storm closed Portland International Airport for a record 2 1/2 days, stranding thousands. Schools closed for three to five days. Ice accumulated to 2 inches in the Willamette Valley. Wind gusts during the storm reached 40 to 50 miles per hour in places.

Elevation: 1,000 to 14,000 feet

Natural vegetation: mixed; broadleaf deciduous and needleleaf evergreen trees

Köppen climate classification: Highland

Most dominant climate feature: prolonged periods of heavy winter rain and snow

By January and February, the wet season in the Cascades is well underway. The region receives as much as 20 inches of rain and snow during this two-month period. The heaviest precipitation drenches the west- and southwest-facing slopes, which take the brunt of the predominant westerly winds associated with powerful Pacific storms. Average daily temperatures range in the 20s and 30s.

Heavy snow and sometimes rain, even at the highest elevations, continue to keep the Cascades wet through March and April. Up to 24 inches of precipitation can fall during March and April alone. The snowpack continues to deepen at higher elevations during March, but wanes by late April. Average daily temperatures moderate slightly and typically reach the 40s by late April.

By May and June, the persistent winter snow and rain begin to decrease, but not before dropping an additional 7 to 14 inches of precipitation. As average daytime temperatures warm into the 50s by June, the heavily monitored snowmelt fills rivers to their banks. The last spring freeze usually occurs during this period, but the highest volcanic peaks can experience freezing temperatures year-round. Though rare, thunderstorms can rumble through the Cascades in May and June.

Summer is in full swing by July, and the region's peak heating and dryness comes in August. The moisture that does fall usually occurs at higher elevations from showers and thunderstorms initiated by orographic lift. Thunderstorms commonly produce gusty winds and dry lightning, which can spark forest fires in dry vegetation. Typically 2 to 5 inches of rain falls in localized storms in the Cascades during July and August. Temperatures maintain comfortable levels in the 70s and 80s during the day and drop into the 50s at night.

The wet season usually begins in late September or in October, totaling as much as 15 inches of precipitation, including the water content in snow, by the end of October. The weather changes dramatically between the first and last part of October: temperatures plummet, snow begins to accumulate, and strong winter winds begin to blow. Average daily temperatures fall from the 50s in September to the 40s in October.

By November and December, the wet season reaches full stride. Precipitation totals for these two months range from 25 to 40 inches. Although most precipitation falls as snow, rainfall sometimes accompanies a warm front. Short-lived, the rain quickly turns to snow that builds some of the world's deepest snowpacks. Temperatures cool into the 20s and 30s during the day, while morning lows dip into the teens and 20s.

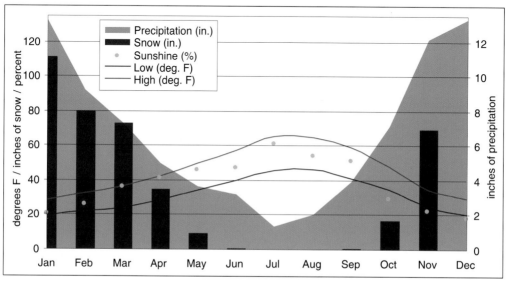

Climate of Stevens Pass, Washington (percent of total possible sunshine estimated)

Rain-on-Snow Events: Great Pacific-Northwest Floods

Rain falling on a snowpack can cause a disastrous amount of flooding in streams and rivers. Although rain itself can melt snow at a rapid rate, a warm, nearly saturated air mass can melt it even faster. Warm air cools as it comes into contact with a cold, snowy surface. This cooling lowers the temperature of the air nearest the snow to the dew point, causing water vapor to condense into water. Since condensation produces a net release of latent heat, the snow warms, which causes fast melting. Sometimes the snowpack in the Pacific Northwest mountains receives both rain and a warm, moist wind, resulting in massive flooding.

With the abundant moisture the Pacific Northwest gets, floods of varying intensity occur relatively often, and people expect them. Several catastrophic floods have swept down the Columbia River Gorge, although increased regulation of the Columbia's flow by hydroelectric dams makes such floods less frequent. During major floods,

The Flood of 1894

One of the greatest Columbia River floods in historic times inundated Umatilla on June 4, 1894. The power of the floodwaters caused the Willamette River to rise to its all-time record stage of 33 feet at Portland, and heavy spring rains coupled with peak flow from snowmelt pushed the Columbia over its banks. The average annual peak discharge of the Columbia River, the largest river in the Pacific Northwest, is about 583,000 cubic feet per second. The flow past The Dalles, Oregon, on June 4, 1894, was an amazing 1.2 million cubic feet per second, swelling the Willamette River and overwhelming downtown Portland's streets with 12 feet of water. Two other great floods in 1862 and 1948 had flows of 1 million cubic feet per second or higher.

water spreads over the flat land on both sides of Bonneville Dam. In extreme floods, roads, highways, and bridges are damaged, but elevated roadbeds and flood control reduce the threat of flooding. When the Columbia River reaches flood stage, 26.5 feet at Vancouver, Washington, its sheer size and volume backs the Willamette River up at the confluence in Portland.

The most deadly flood to sweep down the Columbia River, the famous Vanport Flood, struck in May 1948. No longer a town today, Vanport was a wartime housing project of 18,500 residents, situated near the current location of the Interstate Bridge between Portland and Vancouver—at an elevation lower than the river. A heavy snowpack, combined with above-normal temperatures in late spring and heavy precipitation in the last half of May, sent a tremendous volume of water pouring off the Cascades into streams and rivers to the Columbia River. Water had risen to 26.7 feet at Vancouver on May 29, but then at about 4:00 PM on Memorial Day, May 30, disaster struck the city of Vanport. The water level reached 31 feet—2 feet above the top of the local levees.

Railroad fill—part of the protective dike system around Vanport—failed and allowed floodwaters to inundate Vanport. The prearranged signal for residents sounded, and within minutes a 6-foot-high wall of water swept into the city. The gap in the dike widened to about 600 feet, and two hours later, 10 to 20 feet of debris-laden water overwhelmed the town.

Most people fled Vanport on foot because more than four hundred stalled automobiles jammed traffic along the streets out of town. In only a few hours, the flood left nothing but memories of the city. Thirty-eight people died. However, had the tragedy struck at night as people slept or on a workday when Vanport was filled with workers from outlying areas, loss of life probably would have risen severely. In the greater Portland area the flood destroyed five

Aerial view of the flooded Vanport area west from North Denver Avenue taken after the flood of 1948 —Photo courtesy of the City of Portland Archives and Records

thousand homes, forced one hundred thousand people to evacuate, and left thirty-five thousand residents homeless. According to witnesses, the waters submerged at least 150 cars, trucks, and buses. The National Weather Service ranks the Vanport Flood as Oregon's second most significant weather event of the twentieth century.

The Pacific Northwest's location and the characteristics of the region's storms make it particularly susceptible to heavy rainfall on top of deep snowpacks, known as "rain-on-snow" events. Sometimes Pineapple Expresses unleash these events, providing maximum meteorological melting capability: rain, combined with warm, moist air. The Pacific Ocean supplies storms in this region with abundant moisture, which during the winter can fall as either rain or snow. Even at the highest elevations rain can fall during the winter. Only when cold, arctic air moves into the region will surface temperatures be cool enough to support snow.

After the warm, moist air of a Pacific storm passes through the Pacific Northwest, its counterclockwise circulation draws cold air from Canada into the area. The cold air, coupled with any remaining moisture from the passing storm, helps develop a snowpack. Usually enough cold air from the departing storm remains in the area to keep snow falling until the next moist storm hits the region. Also, during the initial phase

Destruction caused by the Vanport Flood of 1948 —Photo courtesy of the City of Portland Archives and Records

of an incoming storm—before the warm front hits—moist air mixes with arctic air creating more heavy snow, further deepening the snowpack. As the warm front passes over the Pacific Northwest, surface temperatures rise above freezing. This begins a short (usually twelve- to twenty-four-hour) period of rain, even at the highest elevations. The rain continues until the storm's cold front moves across the region and the center of the low moves east, pulling cold air southward from Canada into the Pacific Northwest. As this happens, surface temperatures drop rapidly and change the rain back to snow as the tail end of the storm passes through. Given the right conditions, these storm systems can cause severe flooding.

Other notable rain-on-snow-induced floods swept down the gorge in 1897, 1913, 1916, 1928, 1933, 1950, 1964–65, and 1996. The record flooding in Oregon during late December 1964 and into January 1965 resembled the Vanport scenario of heavy snow followed by persistent heavy rains. The rainstorm of December 20–24, 1964, probably ranks as the most severe to ever hit central Oregon and among the most severe over western Oregon since the late 1870s. Reservoirs filled in early winter, and many could not release water fast enough to prevent overtopping. Water flowing over the top of Dorena Dam, south of Eugene, was more than 8 feet deep. Several weather stations in central Oregon recorded two-thirds of their

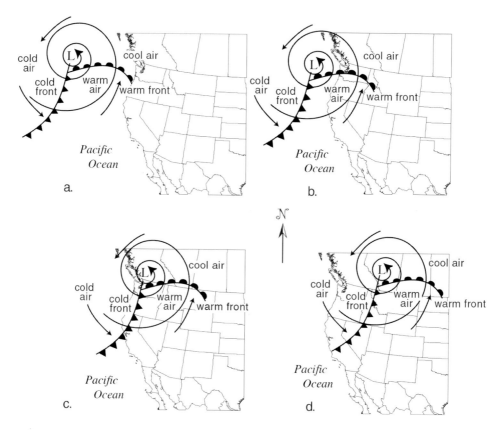

Progression of a Pacific storm as it passes over the Pacific Northwest. (a) The approaching storm spreads rain across Oregon, but cool air to the north supports snow at higher elevations and cold rain at lower elevations. (b) As the storm moves inland, a warm front (associated with warm air pulled in from the tropics) passes from south to north across the region. Snow changes to rain, even at the highest elevations. (c) The cold front associated with cold air pulled in from Canada moves ashore, lowering snow-elevation levels as cold air rushes in from the north. Rain changes to snow at higher elevations, while cold rain occurs elsewhere. (d) Cold, dry air sweeps in from Canada across the entire Pacific Northwest. Although limited at this point, any moisture that does fall, particularly at the higher elevations, will fall as snow.

normal annual rainfall in just five days; normal annual precipitation in this area measures about 50 inches. Several stations set new records for twenty-four-hour totals and December monthly rainfall totals. For instance, Eugene recorded 20.99 inches and Grants Pass, 16.06 inches for the month of December—December records for each location. The Willamette River at downtown Portland reached a flood stage of 29.8 feet, a record high for the winter season and within

inches of the peak stage during the Columbia River spring flood of 1948. The Willamette at Portland normally flows at 7.6 feet.

The 1964 storm triggered widespread severe flooding and mudslides. At least thirty major highway bridges sustained damage so severe that officials closed them. The storm also destroyed the new, multimillion-dollar John Day Bridge and many other bridges on secondary roads. Floods washed out or badly damaged hundreds of miles of roads and highways, destroyed hundreds of homes and buildings, and damaged an even greater number. The flooding claimed seventeen lives and injured scores of others in the Pacific Northwest. The National Weather Service ranks the flood and rainstorm as Oregon's fifth most severe weather event during the twentieth century.

In early February 1996, four days of heavy rain began after a period of extended, bitter cold. Low-elevation snowpacks released up to 10 inches of water in as little as forty-eight hours. Five people drowned and officials declared nearly every county in Oregon a disaster area, including counties in eastern Oregon where heavy rains and frozen soils resulted in flooding as well. Region-wide damage estimates exceeded $1 billion. The flood destroyed hundreds of homes and forced thousands of people to find temporary shelter. The city of Portland erected a makeshift flood barrier to prevent floodwaters from moving into the downtown area. The National Weather Service ranks the February 1996 flood as Oregon's sixth top weather event and the eighth most significant weather event in Washington in the twentieth century.

The Great Spokane Flood

The greatest floods on earth in recent geologic time tore through western Montana, northern Idaho, eastern Washington, and the Columbia River Gorge about 13,000 years ago near the end of the last great Ice Age. Known as the Great Spokane Flood, the Bretz Floods, or the Glacial Lake Missoula Floods, the waters originated in a deep lake that inundated much of western Montana. Tongues of ice from a huge continental glacier dammed the lake until its water level rose enough to float the ice dam like a giant ice cube. Suddenly the ice dam broke loose, sending a towering wall of icy water across what is now Spokane and into the Columbia River. For days, the area that is now Portland, Oregon, lay submerged beneath 400 feet of water. And this happened not just once, but repeatedly—when the lake drained, the glacial ice once again dammed the lake until the water level rose high enough to float the dam and flood once more. The powerful floods deepened and widened the only significant gap in the entire length of the Cascades, the spectacular Columbia River Gorge.

Ninety-Three Feet of Snow: Mount Baker

The northern Cascades are among the snowiest mountains in the world. During the winter months, a continual parade of moist storms moves across the Pacific Ocean toward the Pacific Northwest. The storms slam into the Cascades and are forced upward, where they cool and condense, then drop their heavy loads of moisture. This process of lifting—known as *orographic lifting*—produces orographic precipitation.

The orientation and height of the Cascades make them a very efficient terrain feature for orographic precipitation. Their north-south orientation lies perpendicular to the storm track, so moist storm winds blow up their slopes. The Cascades provide a significant ramp for air to flow up since they rise from about 200 feet to over 14,000 feet. The higher elevations poke into the free atmosphere where air is not warmed by the ocean or affected by Earth's surface friction, so a significant portion of the mountains experience below-freezing temperatures year-round. The mixture of abundant moisture, cold air, and orographic lifting makes the Cascades susceptible to tremendous amounts of snow.

It remains below freezing at elevations over 4,000 feet during the winter months, so snowfall amounts increase rapidly with just small increases in elevation. Some areas receive more than 40 feet of snow in a typical year. However, during the 1998–99 snowfall season, freezing levels remained abnormally low—down to sea level—and it was a La Niña year. La Niña is an

Snowpack, Snowfall, Snow Depth, Snow Base—What Do They All Mean?

Water from snowmelt is among the most vital natural resources in the West. Because of this, the Natural Resources Conservation Service (NRCS) carefully monitors snow. The NRCS operates, installs, and maintains an extensive automated climate station system called SNOTEL (for SNOwpack TELemetry) to collect snowpack and related climatic data in the western United States. Since the 1930s, SNOTEL stations have efficiently collected snowfall data needed to produce water supply forecasts. SNOTEL stations have a pressure-sensing snow pillow that measures the weight of snow, which observers then translate into an estimated snow depth and snow-water equivalent.

Several times each winter, usually on the first and fifteenth of each month, the NRCS does manual snow surveys. NRCS employees set out on skis, snowshoes, or snowcats to measure snow depth, termed *snowpack* in this context, and the snow-water equivalent at specified locations. These measurements provide the most accurate snapshot of the snowpack, while SNOTEL stations provide a good continuous record of snow conditions.

In the context of ski area operators, the term "base" or "snow base" denotes the average snow depth at different elevations of the ski mountain—actually the same measurement as snowpack.

"Snowfall," the total amount of snow that has fallen over a specified period of time, has the same meaning to all people with an interest in tracking the depth of snow.

oceanic weather pattern characterized by unusually cold ocean temperatures in the central and eastern equatorial Pacific, compared to El Niño, which is characterized by unusually warm ocean temperatures in the central and eastern equatorial Pacific. During a La Niña, however, sea surface temperatures usually warm above average over the western Pacific—a breeding ground for subtropical storms that eventually track across the Pacific and into the Pacific Northwest. The warmer sea surface temperatures increase evaporation rates, making more storms for the Pacific Northwest. In 1998–99 La Niña, coupled with abnormally cold temperatures, created an unusually deep snowpack in northern Washington's Cascades.

Mount Baker Ski Area sits at an elevation of 4,300 feet (base elevation at Heather Meadows), 9 miles northeast of the summit of the Mount Baker volcano in northwestern Washington. It receives the highest average annual snowfall of any ski area in North America, with a mean maximum base depth of 186 inches each season. On November 12, 1998, the first major snowstorm of 1998–99 blanketed the ski area with 33 inches of snow over a five-day period. This snowstorm—and many others that season—buried the remote ski area. The biggest twenty-four-hour snowfall of the year for the ski area during the 1998–99 season arrived on December 2, 1998, when 36 inches of snow fell.

Months into the snowfall season, on February 22, 1999, a headline in the *Seattle Post-Intelligencer* read, "Rare ski complaint: Too much snow." By this time, 19.5 feet of snow had fallen on the Mount Baker Ski Area since the first of February, and another winter storm was blowing in. With more snow than the ski resort staff could handle, the ski area closed for two days to give crews a chance to dig out parking lots, ski lifts, and buildings. The snowbanks along the road grew so high that front-end loaders, D-9 Caterpillar bulldozers, snowblowers, and plows had to cut them down so they would not tunnel over the road. Avalanche control teams reported using 200 percent more dynamite than usual to reduce avalanche potential, and goggle sales hit a record high. In February 1999 alone, 304 inches of snow fell at Mount Baker Ski Area, topping off an amazing and dangerous winter.

By May 12, 1999, the seasonal snowfall total at the ski area had reached 1,124 inches—a total snow depth equaling the height of a thirteen-story building. This amount surpassed the old world record of 1,122.5 inches of snow recorded at Paradise, Washington, on the slopes of Mount Rainier during the snowfall season of 1971–72. By the time meteorologists tallied the year's

More Elevation, More Snow

On the western slopes of the Cascades, the slightest increase in elevation profoundly increases the amount of snowfall a location receives. The farther storms travel up the mountains, the colder it gets, causing more of a storm's moisture to condense and fall as snow. For example, Seattle receives an average 13 inches of snow each year while atop Snoqualmie Pass, elevation 3,022 feet, 440.4 inches of snow falls annually—an increase in annual average snowfall of 17.1 inches for each 100-foot gain in elevation.

You may wonder if the amount of precipitation the lower elevations receive equals the amount of snow in higher elevations if it is melted down? No. The higher elevations actually receive *more* precipitation because of the intense orographic lifting. For example, between Phoenix and Flagstaff, Arizona, the snowfall increases from no snow in Phoenix to 100 inches in Flagstaff, which is nearly 6,000 feet higher—that's a rate of 1.7 inches for every 100 feet of elevation gained—a mere shadow of the remarkable effect of elevation on snowfall in western Washington.

snowfall statistics from July 1, 1998, through June 30, 1999, they totaled 1,140 inches—a new world-record snowfall measured over a twelve-month period, a record that still stands today. The National Climate Extremes Committee, which evaluates potential record-setting weather events in the United States, scrutinized the snowfall figures and made the record official on August 3, 1999. The deepest snow base of the 1998–99 season—318 inches—broke the old record at Mount Baker of 300 inches measured at Heather Meadows in April 1946.

Snow can fall at any time of the year in most mountains of the West, including the Cascades, but the snowiest period runs from October through April, when the storm track, known as the *jet stream,* positions itself directly over the Pacific Northwest. Storms hit about every other day during winter and can deliver phenomenal amounts of heavy, wet snow. Because of its proximity to the Pacific Ocean's damp air masses, the Cascades collect some of the world's heaviest snows, which feed some of the lower forty-eight's only active glaciers.

Snowfall Totals at Mount Baker Ski Area, Washington, July 1, 1998–June 30, 1999

MONTH	AMOUNT (inches)
July–October 1998	178
November 1998	225
December 1998	182.5
January 1999	303
February 1999	194
March 1999	24.5
April 1999	30
May–June 1999	3
Total	1,140 (95 feet)

An A-frame building and truck at the Mount Baker Ski Area before the snowfall of the 1998–99 season and after nearly 95 feet of snow had fallen. The winter photo was taken in May 1999 after about 10 feet of snow had already melted. —Gwyn Howat photos, Mount Baker Ski Area

Mount Hood capped by a lenticular cloud, photographed from an airplane on the morning of March 11, 2004 —Phil Pasteris photo, Natural Resources Conservation Service

The southern Cascades, like the northern Cascades, also experience amazing snowstorms. Mount Shasta Ski Bowl, which sits east of Mount Shasta, California, holds the national all-time record for the greatest snowfall from a single storm.

Soggy Snow

Because of its proximity to the Pacific Ocean, some of the world's wettest and sloppiest snow flies atop the picturesque Cascades. A meteorological rule of thumb holds that 10 inches of wet snow melts down to 1 inch of water—a 10-to-1 ratio of snow to water. However, at Snoqualmie and Stevens Passes in Washington, an early winter or spring snowstorm can drop 20 inches of snow that melts to as much as 4 inches of water—a 5-to-1 ratio, twice the norm. Take, for example, the soggy day of January 9, 1990, when 22 inches of wet snow fell at Stevens Pass, Washington. That snow equaled 4.19 inches of liquid precipitation, a snow-to-water ratio of 5 to 1, and a daily precipitation record for the region. That's wet snow.

Needless to say, the Cascades, unlike Utah, are not known for their dry, powdery snow. To put the soggy Cascades snow in perspective, sometimes 40 or even 50 inches of Utah's snow yields a single inch of water—a 40-to-1 or 50-to-1 ratio of snow to water. But while skiers in Washington or Oregon may splash when they fall down, the deep snow packs make it possible to ski almost year-round. Timberline Ski Area, perched on Oregon's Mount Hood, has the longest ski season in the lower forty-eight. Typically, it officially opens in September and closes in July. With an average annual snowfall that exceeds 400 inches and a few permanent snowfields, the area draws the U.S. Olympic ski team for summer training.

Parades of moist Pacific storms march across the state each winter. As the jet stream hurls storms over the Cascades, the mountains wring out the moisture, creating intense periods of heavy snow. Snowfall rates of 4 to 6 inches per hour occur commonly during the height of an extreme snow shower, restricting visibility to zero. Lightning and thunder, often associated with such heavy snow showers, indicate a great deal of uplift in the atmosphere and a great deal of instability.

The unprecedented Mount Shasta Ski Bowl storm of 1959 began dropping snowflakes on February 12. By February 13, the snow fell more steadily; dropping 2 inches by day's end. And it snowed, and it snowed, and it snowed for seven days straight at an average rate of 1 inch per hour; on February 16 and 17 the snow fell at the rate of 2.2 inches per hour. By February 19, the area's busy weather observer had measured a whopping 189 inches (15.8 feet) of snow and 14.36 inches of precipitation or melted snow. It snowed an average 1.1 inches per hour for 168 hours to total an average 2.25 feet per day! By storm's end, the snow depth at the base of Mount Shasta Ski Bowl tallied 247 inches, making this snowstorm the heaviest snowstorm in U.S. history—it dropped the most snow ever recorded from a single storm.

The only storms that challenge the intense snowfall associated with the 1959 Mount Shasta storm in the Cascades have struck the Sierra Nevada. In January 1952, a six-day storm dropped 149 inches (12.4 feet) of snow at a Lake Tahoe, California, ski resort. Similarly, a December 20–23, 1996, storm buried the Alpine, California, area in 7 feet of snow in just three days—4 feet of that came in just twenty-four hours. Snowplows could not keep up with the heavy snow, so officials closed Interstate 80 for nearly two days, as they commonly do during severe winter storms. This event approached but didn't break the state twenty-four-hour snowfall record of 60 inches set at Giant Forest, California, on January 19,

1933. The country's twenty-four-hour record of 75 inches fell at Silver Lake, Colorado.

The Wellington Avalanche: Stevens Pass, Washington

The steep gradients in the Cascades and heavy snowfalls leave slopes along the eastern and western valley walls vulnerable to avalanches. Since precipitation typically switches back and forth between rain and snow in this region, icy slabs develop between storms. These icy layers provide an optimal medium for avalanches to slide on. If several feet of fresh snow accumulates on an old rain-crusted slab, the newly deposited snow becomes increasingly unstable. Despite an increased number of backcountry users in the last twenty years, avalanche deaths in the Northwest, according to the Northwest Weather and Avalanche Center, have dropped from an average of three per year to about one based on a five-year average. This decrease comes from increased awareness and preparedness and better forecasts. Even so, avalanches are common in the Cascades, and this region of the United States has given history some of its most notorious avalanche tales.

On the morning on February 23, 1910, two trains left Spokane, Washington, at different times and headed for depots along the Puget Sound. Train number 27 was the Great Northern's *Fast Mail,* and train number 25 was a twelve-wheeled passenger train with thirty-eight men, nine women, and eight children on board. After passing through the rolling hills of eastern Washington, they both eased to a halt in Leavenworth, at the foot of the Cascades, and waited for the weather to improve. After a short delay it stopped snowing, and the trains continued up the east side of the Cascades, only now with rotary snowplows ahead of them to clear away the 8 to 9 feet of snow already on the ground. A storm of huge proportions threatened to bear down on the region.

After the trains passed through the Cascade Tunnel east of Stevens Pass (4,061 feet), the superintendent of the Cascade train routes, James H. O'Neill, ordered the trains to stop at Wellington. One-quarter mile west of the tunnel, a cluster of buildings, railroad sheds, and bunkhouses on a triangular mountainside shelf made up the town of Wellington. Surrounded by huge snowdrifts, the trains sat side by side, perched on the slope of Windy Mountain. Passengers accepted an invitation to eat at the hotel as small avalanches raced down the mountainside. In his two-and-a-half years on the job O'Neill noted, he had put more than four thousand trains through the mountains in snowstorms, and before February 23, 1910, none of his trains had been delayed more than twenty-four hours. In the years since regional railroad service had begun in 1893, railroad workers had learned that such fierce, late-season snowstorms did strike the Cascades, but usually let up after a day or two. O'Neill felt that by the next day, February 24, the weather would ease.

But the snowstorm raged on. Some passengers claimed snow fell at the rate of 1 foot per hour—extreme, to be sure, but possible. For days O'Neill's crews fought to clear snowslides off the tracks. On their third day on the mountain, passengers grew edgier, and several demanded that the passenger train be backed into the Cascade Tunnel to move it away from the threatening snowslides. The tunnel was cold and not adequately ventilated, and O'Neill didn't want to risk having his passengers asphyxiated aboard a running train, so the train stayed next to number 27 out in the open.

Through two more tense days and nights, the forceful storm continued. On the night of February 28, the falling snow changed to rain, and lightning crackled through the Cascades. Such a change from snow to rain is common for winter storms in the Cascades. Initially, the Cascades are positioned ahead of a warm front of a storm, so temperatures remain below freezing. As the warm front passes (usually from the southwest

CUTTHROUGH SNOW SLIDE AT WELLINGTON   MARCH 1910

Men stand on the railroad tracks after snow from the avalanche had been removed near the Wellington Train Station. It took twelve days to clear the tracks. —J. D. Wheeler photo, courtesy of Warren Wing

to the northeast) over the Cascades, temperatures rise and force the precipitation to fall as rain. The short-lived rain usually lasts twelve to twenty-four hours. As a storm's associated cold front sweeps westward across the Cascades, temperatures plummet, and the rain quickly changes back to snow, but usually not for long since the main storm system has moved east of the Cascades (see *Rain-on-Snow Events: Great Pacific-Northwest Floods* in this chapter).

At 1:20 AM on March 1—the exact time determined by the stopped watches on bodies of the victims—a snow mass roughly 500 feet above the trains broke loose along a line as neat as if it had been snipped by giant shears. Avalanche experts think rising temperatures lubricated the great mass of snow and ice, and possibly a stroke of lightning jarred it loose. About 3,000 feet wide, the avalanche rushed down the slope. A 15-foot-

deep slab of snow came off the mountain, but as the avalanche plowed down the hillside it grew to a depth of 40 feet at its base. Acres of snow roared over Wellington Station, gathering everything in its path. The massive slide instantly swept both trains sideways over the mountain ledge and deposited them 150 feet below in the Tye Valley. Along with the trains, the avalanche demolished a 300-foot-long building and some railroad shacks. One hundred and one passengers and crew lost their lives. Fortunately the avalanche missed most of the station's buildings, bunkhouses, and hotel, which had filled with people.

Help arrived slowly because the unrelenting storm had buried the Cascades. The tracks west of Wellington finally opened on March 9—two weeks after the storm began—when a rotary plow broke through from the east. And finally, on March 12, workers using blasting powder forced

Locomotives for the passenger and mail trains lie side by side at the Wellington Station in Washington. —J. D. Wheeler photo, courtesy of Warren Wing

open the route to the west, which was clogged with numerous other slides—eighteen days after O'Neill first sent a plow into the snow at Windy Mountain.

The storm had brought ten solid days of accumulating snow and ice. The rescue effort

Deadly Utah Avalanche

Another avalanche catastrophe virtually destroyed the small mining town of Alta, Utah, in February 1885. However, Utah's most deadly avalanche occurred on February 3, 1926, when a huge snowslide in Bingham Canyon demolished fourteen miners' cottages and a three-story building. The slide killed thirty-six people and injured another thirteen of the sixty-five in its path. The National Weather Service ranks the event as Utah's third most significant weather event of the twentieth century.

was so difficult, and the snow packed so hard into the valley bottom, that it took the arrival of spring when enough snow had melted for rescuers to recover the first body. To blot out the memory of the wreck, Great Northern changed the name of the station from Wellington to Tye and years later abandoned the tunnel and station altogether. A new, better, and safer tunnel, almost 8 miles long, was built by Great Northern farther down the slope.

On a personal note, one of my ancestors, John K. Parzybok, died in the accident. Part of the crew, Parzybok lies with his coworkers in Everett, Washington.

Extraordinary Precipitation Gradient: North-Central Oregon

Most people think of Oregon as a place of plentiful rainfall, thick forests, and waterfalls. But that description fits only the western one-third of the state. The eastern two-thirds of Oregon lies in the rain shadow of the Cascades—among the most effective precipitation barriers in the West.

The average annual precipitation gradient—that is, the change in average annual precipitation over a horizontal distance—across the crest of the Cascades ranks among the most abrupt in the world. The western slopes of the Cascades experience a moist, marine climate and receive abundant moisture. In contrast, dominated by a high-desert climate, the eastern slopes lie in a rain shadow and receive far less moisture.

Orographic lifting—the process by which air rises up mountain slopes, cools as it rises, and forms clouds and precipitation—wrings moisture-laden Pacific storms dry, leaving little moisture to dampen thirsty soils of eastern Oregon and Washington. Although some moisture blows over the crest of the range and falls just east of the ridgeline, downslope winds—tail ends of the Pacific storms—quickly dry through compressional heating, eliminating any chance for rain- or snow-producing clouds to form. Satellite images show cloud cover ending immediately at the crest of the Cascades, leaving eastern areas under clear skies. Likewise, radar imagery shows precipitation ending at the ridgeline.

The most sudden change in average annual precipitation occurs near Santiam Pass, Oregon, along U.S. 20, about 90 miles south of Hood River. The average annual precipitation drops from 90 inches near the summit of Santiam Pass, to 14 inches just 20 miles east in Sisters, Oregon (elevation 3,186 feet)—a change of 3.8 inches of precipitation per mile. Although the horizontal gradient (per mile of distance) may amaze the casual observer, the vertical gradient (per foot of elevation) proves even more significant. Santiam Pass, 4,817 feet above sea level, and Sisters, at 3,180 feet, differ a mere 1,570 feet, but Oregon residents in this region see a staggering change in average annual precipitation of .05 inch per foot of elevation!

The precipitation gradient through the Columbia River Gorge, while not as extreme, is remarkable as well. Within the 25-mile stretch of Interstate 84 from The Dalles to the Bonneville Dam, the average annual precipitation increases from a meager 20 inches to nearly 80 inches. In other words, for every mile west of The Dalles, average annual precipitation increases 2.4 inches. The Dalles resides only 10 feet higher than Bonneville Dam, so the rate of change per foot of elevation does not apply. Because the gorge cuts such a narrow path through the Cascades, the

The Driest Place in Oregon

Oregon's unpopulated Alvord Desert, located in the great rain shadow of the Cascades and in the local shadow of Steens Mountain in southeastern Oregon, can lay claim to the title of the driest spot in the state. While Steens Mountain receives upwards of 52 inches of precipitation a year, the Alvord Desert, on the eastern side of Steens Mountain, receives a mere 7 inches—that's a difference of 45 inches within a distance of about 5 miles.

Steens Mountain, which rises to 9,733 feet, wrings out any remaining moisture from the few clouds that make it east across the Cascades. Therefore, little if any moisture makes it to the Alvord Desert from the Pacific. The primary precipitation source for this desert comes from scattered summer thunderstorms tapping monsoon moisture traveling from the south in the summer.

While Pacific storms annually deliver nearly 200 inches of precipitation to the lush northwestern corner of Oregon, the western mountains cast a profound rain shadow over eastern Oregon—a rain shadow creating a difference in precipitation of 190 inches across the state.

overwhelming rain shadow effect created by the surrounding mountains causes the gently sloping gorge to sit in a rain shadow as well.

A few areas along the crest of California's Sierra Nevada and the northeastern reaches of Washington's Olympic Mountains have precipita-

tion gradients similar to, but less extreme, than those in the Cascade Range. The next time you travel across the Cascades, witness the striking changes in the general landscape, plant species, and vegetation density. These changes result largely from the abrupt difference in precipitation.

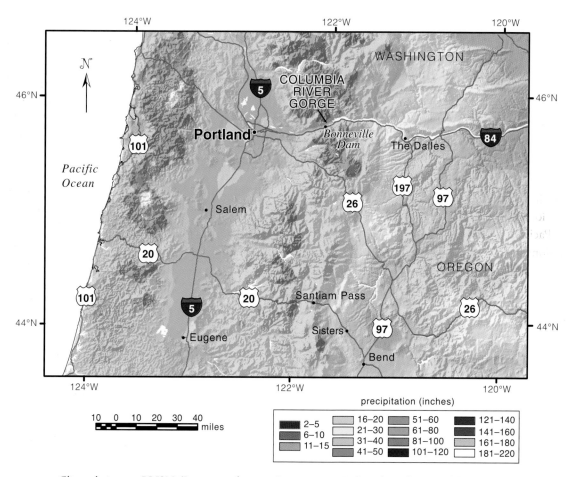

Climatologists use PRISM (Parameter-elevation Regressions on Independent Slopes Model) to interpolate average (1961–1990) annual precipitation in northwestern Oregon. The precipitation gradients between The Dalles and the Bonneville Dam as well as across the crest of the Cascades near Santiam Pass are clearly visible. —Oregon State University/Oregon Climate Service

3

The Pacific Northwest's
Inland Empire and High Deserts

Elevation: 300 to 9,600 feet

Natural vegetation: short-grass prairie (steppe) and semiarid-desert plants

Köppen climate classification: steppe and semiarid

Most dominant climate feature: in the rain shadow of the Cascades, generally dry and protected from strong Pacific storms

The sun's low angle and the frequent snow cover make January and February the coldest months of the year in this region. Typically, daytime temperatures reach only the 30s, and morning lows drop into the teens and 20s. The Cascades weaken strong westerly Pacific storms, so the inland Northwest receives much lower amounts of precipitation than western Oregon and Washington. Typically 2 to 5 inches of precipitation including melted snow falls in this area in January and February, mainly as snow or cold rain. Frequent episodes of fog and low clouds can keep the lowest elevations cold and frosty for days, while the high deserts generally receive more sunshine. The desert's clear skies, however, allow for extremely cold nights; double-digit subzero temperatures can sometimes linger for days.

March, relatively wet throughout this region, contributes 1 to 3 inches of precipitation to annual totals; things begin to dry out a bit in April, when an average ½ to 2 inches of precipitation fall. Most of this moisture falls as snow, but rain becomes more common as spring progresses.

Average daily temperatures moderate to the 50s by late April. Most low-elevation areas experience their last spring freeze in late April, while the high deserts can have freezes year-round.

May and June mark a welcome end to winter's cold and foggy grip. Temperatures warm to the 70s and 80s during the day and fall to comfortable 40s and 50s at night. With spring comes the first of this area's annual ten to twelve thunderstorms, sometimes accompanied by gusty winds, small hail, and frequent lightning. The combination of these storms and other showers brings 1 to 4 inches of precipitation to the region. Later in the summer, strong winds associated with an exchange of air through the Columbia River Gorge often contribute blowing dust and dust devils.

Peak heating and dryness occur in July. Although daytime temperatures typically warm into the 80s and 90s, 100°F heat occurs in the hottest areas near the Columbia River during July and August. Nighttime lows reach the 40s in the high-elevation areas of Oregon and the 60s in the warmest areas of east-central Washington. Occasional showers and thunderstorms bring welcome but light precipitation. Precipitation totals fall typically between ¼ and 1 inch for July and August.

As in most of the Pacific Northwest, September arrives mild and dry in the inland Northwest, and October quickly turns colder and moister with the onset of the wet season. Most

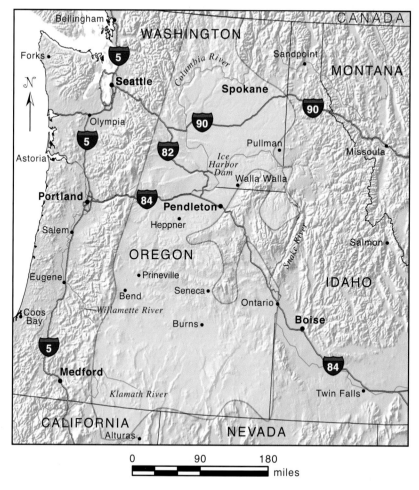

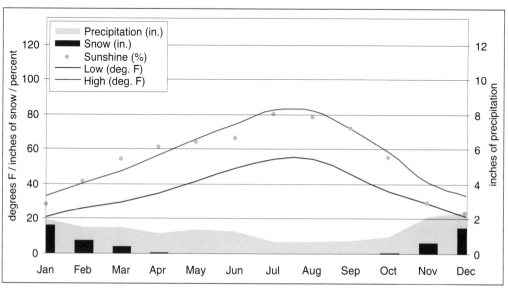

Climate of Spokane, Washington

Mammatus clouds hang beneath a rare and severe thunderstorm east of Spokane, Washington, during summer 1999. —Michael Jewell photo

areas experience their first freeze and their first snowfall during September or October. During this two-month period, 1 to locally 4 inches of precipitation normally dampens this region, most of it coming in October. Average daily temperatures cool from the 50s and 60s in September to the 40s and 50s in October.

Temperatures and precipitation continue to fall in November and December. By December, high temperatures range from the 30s and 40s in the elevated terrain of Oregon and north-central Washington to the 40s elsewhere. One to 4 inches of precipitation accumulates, falling as snow at the highest elevations and a mix of rain and snow at the lower elevations. Ice storms dominate during this time of year, particularly along and near the Columbia River and across eastern Washington.

Dust Devils: Eastern Washington

Guaranteed—you will see a dust devil in eastern Washington on almost any hot, sunny summer afternoon. The dry, cultivated terrain and intense local heating make this area ideal for the development of the world's smallest storm, the dust devil.

Dust devils, also called whirlwinds, form under clear skies in areas that generally have low humidity. The dry air doesn't form clouds, which can trap heat, so nighttime air temperatures tend to be cooler. During the day, the sun heats the earth's surface and the air immediately above it. A thermal gradient develops between the warm surface air and the cooler air above, and a column of rising superheated air develops within a chimney of relatively cooler air. The larger the thermal gradient, the faster the heat from the surface will rise. As the warm air rises, a void of air forms at the surface then fills with hot air and feeds the chimney. Since rising air is associated with low-pressure systems, the air begins to rotate counterclockwise as it rises. Although not necessary for dust devils, hilly terrain—like that found in eastern Washington—enhances the spiraling motion as the counterclockwise winds are deflected off the hills. Terrain can even cause curling winds to rotate clockwise. Once

A strong June dust devil whirls near Hatton, Washington.

the whirling column of air can entrain sand and dust, it becomes a bona fide dust devil.

Typically erratic, dust devils usually rotate counterclockwise. The devils can skirt across the landscape at a speed of up to 15 miles per hour, or they can remain stationary. Most of the smaller dust devils last a minute or less, but larger ones can swirl for several hours.

Dust devils can vary considerably in size. Diameters range from 10 feet to more than 100 feet, and heights range from 10 feet to more than 5,000 feet in extreme cases. One of the world's largest documented dust devils developed over the Bonneville Salt Flats in Utah. At one point it extended 24,500 feet into the atmosphere and traveled more than 40 miles.

Snow Devils

Vortexes—tightly rotating columns of air—that form over snow-covered terrain are called *snow devils* or *snow spouts*. Meteorologists have just begun to study this form of whirlwind. Snow devils, by-products of wind eddies, form as winds curl around hills or mountains. Surface heating does not affect the development of snow devils as it does with dust devils.

Wind speeds within whirlwinds commonly reach 15 to 35 miles per hour and in some cases exceed 50 miles per hour. Upward air currents can reach 25 to 30 miles per hour, powerful enough to pick up and hurl fair-sized debris, such as sand, tumbleweeds, and trash. Some severe dust devils have destroyed barns and mobile homes, blown trucks off highways, and hurled up to 50 tons of topsoil into the sky.

While no statistics exist on the number of dust devils spinning across eastern Washington each year, the total could easily reach the tens of thousands. Eastern Washington is by no means the only place in the country where dust devils stir up the landscape. These vortexes occur anywhere intense surface heating occurs, but only become visible when they can entrain debris. Areas of southern Arizona, western Colorado, southeastern California, New Mexico, and the open fields of Iowa and Wisconsin are especially prone to violent dust devils because they have open landscapes and experience intense surface heating.

The Heppner Disaster: Oregon

At about 5:30 PM Sunday evening, June 14, 1903, a torrential thunderstorm unleashed heavy rainfall and hail over the hills south of Heppner in north-central Oregon. At the time about 1,400 people lived in the small farming community of Heppner, nestled along Willow Creek. As the rain fell, Skinners Fork and other small tributaries of Willow Creek east of town rose steadily, and huge

Destruction in downtown Heppner, Oregon, following the June 14, 1903, flood —Photo courtesy of U.S. Army Corps of Engineers, Portland District

amounts of water began heading toward town. Picking up debris from area farms, floodwaters began to fill Willow Creek like never before. The ensuing flood became known as the "Heppner Disaster."

The meteorological conditions of the storm suggest it was a cloudburst—intense and highly localized rain associated with a thunderstorm—somewhat common over this area. In fact, the storm presented no startling features, except the amount of hail it dropped. John T. Whistler, an agent of the U.S. Geological Survey, visited Heppner shortly after the flood. His report suggested the storm extended from the foot of the Blue Mountains, east of town, 8 to 10 miles across the Willow Creek valley and covered a strip only 2 to 4 miles wide. A "phenomenal fall of hail" preceded the rainfall, with some hailstones measuring up to 1¼ inches in diameter. In fact, evidence suggested that several people died from hail. Storms that can drop several inches of hail rarely occur in eastern Oregon.

More often than not, moisture is the missing ingredient in thunderstorms across this arid land-scape. Although not common, flood-producing thunderstorms do occur when monsoonal moisture is drawn north into Oregon (see *Southwest Monsoon* in *chapter 6*). When coupled with intense surface heating, moisture rises, cools, and condenses into cloud droplets. Eventually a mature thunderstorm forms. Since the polar jet stream sits well to the north, and the subtropical jet stream lies well to the south when the monsoon occurs, weak upper-level winds prevent the thunderstorm from moving very quickly; therefore, these storms can drop heavy rains over a concentrated area.

At the southern end of Heppner, a large steam laundry facility stretched across Willow Creek. The floodwaters deposited debris against the facility's abutments until the debris effectively dammed the water. The river pooled but the building could not withstand the tremendous pressure, and it collapsed. The resulting wall of water hit Heppner with a massive force. The most conservative eyewitness reports estimated the wall of water reached 15 feet high, enough to flood homes to knee depth on the second floor.

Within minutes, the flash flood swept away one-third of Heppner. The town suffered $250,000 ($4.9 million in 2002 dollars) in property damage. More tragic, though, hundreds died in the

Hail Damage in Hermiston

On July 9, 1995, severe thunderstorms and flash flooding struck central and northeastern Oregon. Though somewhat rare in Oregon, a supercell thunderstorm developed near Redmond and traveled nearly 200 miles before dissipating. It pelted cities from Condon to Hermiston with baseball-sized hail. The hail damaged nearly every vehicle in Hermiston. The local watermelon crop, ready for harvest, was a complete loss. The storm spawned flash floods, damaging winds, and even a brief tornado. The National Weather Service's new Doppler radar tracked the storm and allowed forecasters to provide citizens ample warning. No fatalities resulted, but damages to crops, structures, and property amounted to tens of millions of dollars. The National Weather Service ranks the event as the seventh most significant weather event in the twentieth century in Oregon.

disaster. A temporary morgue was set up in the Roberts Building, one of the few structures still standing on Main Street after the flood. Rescue workers dug some of the bodies from the debris several miles downstream. They found others fully preserved in huge drifts of hail. Some bodies weren't recovered for five days, and some people were never found. The official count: "approximately 250 dead."

The citizens of Ione, just 20 miles downstream of Willow Creek, would have met a similar fate if not for telephoned warnings. The alerts prompted officials to evacuate Ione immediately, and residents escaped to high ground. At least 150 homes were destroyed in Ione, and the bodies of some unlucky Ione residents washed more than 20 miles downstream to the Columbia River.

The flash flood still stands as one of Oregon's deadliest weather disasters. Although the thunderstorm itself was not unusual, the ruggedness of the topography, coupled with heavy rain and hail, produced the deadly flood. As damaging as the event was, Willow Creek returned to its normal flow an hour and a half after the crest of the flood. The Heppner Flood was as deadly as the famous Big Thompson Flood in Colorado and the Rapid City Flood in South Dakota (discussed in later chapters). The Portland office of the National

Central Heppner, Oregon, on June 14, 1903, following the Heppner Disaster —Photo courtesy of U.S. Army Corps of Engineers, Portland District

Weather Service ranks the Heppner flood as the number one weather event of the twentieth century in Oregon.

Oregon and Washington Hot Spots

East of the Cascades in Oregon and Washington the land becomes flat to gently rolling, sparsely vegetated semidesert. Steep canyon walls rise along rivers and creeks. The bedrock is basalt, and sagebrush dominates the vegetation throughout much of the region. The stretch between The Dalles and Pendleton is the hottest area of Oregon. Likewise, the adjacent area across the Columbia River in Washington is among the hottest in that state. The topography is largely the reason these areas are notorious for their heat.

Lying in the rain shadow of the Cascades, this region receives only 8 to 12 inches of precipitation a year. Without the moderating effects of moist air, the region experiences cold and hot temperature extremes. Limited precipitation limits the amount of evapotranspiration—the combined loss of water through plant transpiration and the evaporation of water from soils and water bodies. Plants transpire water through pores as part of their natural metabolic process. As water evaporates from a plant, it draws heat, cooling the air in the process. Evaporation also cools the surrounding air. Given this region's sparse vegetation, more of the sun's energy directly warms the earth's surface, which then conducts heat to the air.

At an elevation of 300 to 1,000 feet above sea level, the area along the Columbia east of the Cascades is some of the lowest terrain in Oregon and Washington. Lower elevations have higher air pressure and higher air density than higher elevations, and both of these factors contribute to warmer air temperatures. The prevailing westerly winds across this area descend the Cascades, undergoing adiabatic warming and further heating air temperatures. Consider a parcel of air at some high elevation. If the parcel moves to a lower elevation, the air pressure increases with decreasing

elevation, forcing the parcel to compress. This compression causes air molecules within the parcel to gain kinetic, or internal, energy, and to move faster. As the air molecules speed up, the temperature increases within the parcel. All of these processes combine to make the semidesert east of the Cascades very hot at times.

Record-breaking heat scorched two different Oregon towns within a week in 1898. On July 29 a sizzling heat wave drove the temperature to 119°F—the hottest temperature ever measured in Oregon—at Prineville. Just a week later, on August 10, the mercury rose to 119°F in Pendleton. The hottest temperature ever recorded in Washington, 118°F on August 5, 1961, humorously occurred at Ice Harbor Dam on the Snake River 30 miles north of Pendleton. That high tied the previous record of 118°F recorded on July 24, 1928, at Wahluke, Washington, which sits on the southern slopes of the Saddle Mountains in Grant County, northeast of Yakima.

The low elevations of central and south-central Washington typically record the hottest temperatures of the state on any given summer day. The average high temperature in July in this area reaches a blistering 90°F. In Oregon, the upper Columbia Basin, the Rogue Valley, and northern Malheur County host the three hot spots. In terms of the warmest summers, the town of Medford is the warmest, with its July high temperature averaging 90.5°F, but places like The Dalles and Pendleton follow close behind. Medford experiences warm winters because of its low elevation (1,382 feet) and because the mountains surrounding the city protect it from cold air intrusions from the north. In both states, the warmest place in the winter is on the west side of the Cascades, where the relatively warm marine air moderates temperatures.

Oregon's Icebox: Seneca

In early February 1933, record cold temperatures (–90°F) gripped Siberia while Oregon basked in relatively mild weather. Strong westerly winds pushed the frigid air mass from Siberia toward Alaska, setting record lows in the Last Frontier. Then an area of high pressure began to build in the Gulf of Alaska. The clockwise wind circulation around this feature forced the cold air to flow southward across western Canada and directly into eastern Oregon. It eventually reached all the way to the Gulf of Mexico. The National Weather Service ranks the severe but short-lived 1933 arctic outbreak as the tenth top weather event in Oregon during the twentieth century.

On the morning of February 10, the temperature plummeted to –54°F at Seneca, in east-central Oregon, compared to the preceding daytime temperatures in the teens. Just a day before, the temperature at Ukiah, Oregon—a nearby station in a similar high valley—dropped to –54°F, establishing a new record for the state (the Seneca reading stands as the official coldest temperature ever recorded in Oregon because it represents the most recent recording of that temperature). Almost as remarkable as the –54°F temperature at

Seneca was the phenomenal warming afterward. The next day, February 11, the high temperature warmed almost 100°F to 45°F!

Since 1933, the temperature at Seneca has approached –54°F a number of times, including two relatively recent recordings: –48°F on December 23, 1983, and February 6, 1989. According to unofficial reports before the Seneca station existed, temperatures in the area dropped to –60°F during the so-called Big Cold Snap of 1927. So what makes Seneca, Oregon, such a cold spot?

Seneca sits at 4,700 feet in the deep Bear Valley, 20 miles south of John Day, Oregon. Mountains more than 7,000 feet in elevation completely surround the town. At night cold, dense air from atop the mountains to the north settles in the valley and becomes trapped. Narrow valleys that frequently trap cold air are known as *frost hollows* or *frost pockets*. The cold air in frost pockets and frost hollows makes it easier for frost to form than in nearby higher terrain. During the day sunshine heats the surface of the mountains, causing the air to warm more than the air trapped

Temperature Table for Seneca, Oregon, Based on Data Collected between 1949 and 2003

MONTH	MAXIMUM (average)	MINIMUM (average)	AVERAGE	MINIMUM (day/year)
January	34.2	10.1	22.2	–43 (26/1957)
February	39.3	13.8	26.6	–48 (06/1989)
March	44.8	19.7	32.2	–25 (05/1955)
April	53.4	25.5	39.4	5 (09/1952)
May	61.7	31.2	46.5	11 (01/1953)
June	70.6	35.9	53.3	17 (04/1962)
July	80.8	37.6	59.2	19 (08/1959)
August	80.6	35.0	57.8	14 (24/1992)
September	71.8	28.0	49.9	5 (30/1954)
October	59.8	21.5	40.7	–11 (31/2002)
November	44.1	18.6	31.3	–31 (15/1955)
December	35.5	12.4	23.9	–48 (23/1983)

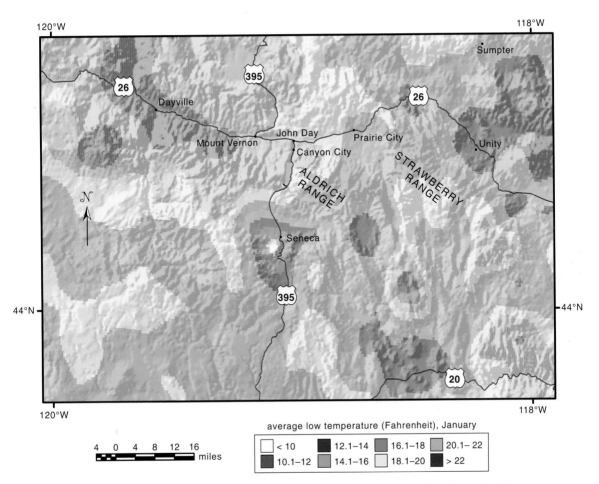

Map of average February low tempeature (based on data collected between 1961 and 1990) of the area around Seneca, Oregon. Seneca is surrounded by mountains, including the Aldrich and Strawberry Ranges. This map was created using Oregon State University's Spatial Climate Analysis Service PRISM.

in the valley bottoms, where the formation of fog also prevents surface heating. The relatively warm layer of air over the valley causes an inversion that acts as a lid, preventing the cold, dense air from escaping or mixing with the air above.

During a strong arctic cold snap, the temperature at Seneca can linger below zero for days and even weeks. January 1, 1986, marked the end of a twenty-one-day streak of consecutive below-zero days at Seneca, the longest such streak at Seneca since digitized records began in 1948

(older records exist, but they have not yet been keyed into a digital database).

A frost pocket illustrates a classic example of a microclimate—a variation of climate in a small area. Since Seneca lies in a frost pocket, it frequently reports colder temperatures than other sites in eastern Oregon. Even Seneca's summertime lows can fall below freezing! Frost pockets commonly occur elsewhere in the Inland Empire and around the world, but Seneca ranks among the most well known.

The Inland Northwest's Worst Winter Storm: Spokane

On the eve of November 18, 1996, the perfect mix of weather ingredients delivered the inland Northwest's worst ice storm in sixty years to the Spokane area. It caused the worst weather-related power outage in 108 years. The ice storm began as a record-breaking rainstorm in Hawai'i several days earlier. The storm hit the island of Oahu particularly hard, dumping nearly 20 inches of rain, which resulted in extensive flooding. During a month that normally brings 3 inches of rain, a monthly record 18.70 inches soaked Honolulu. A series of storms along a jet stream plume of moisture dropped the surprising amount. And when that moist storm took aim at the Pacific Northwest, it gave Oregon its wettest day of the twentieth century: Port Orford, Oregon, received a state-record 11.65 inches of rain in a twenty-four-hour period between November 18 and 19, 1996.

The moisture plume—a Pineapple Express (see *chapter 1*)—left Hawai'i and moved across

Rime ice coats trees in Spokane, Washington, on February 2, 1997.

the Pacific into Oregon and Washington, and the head of the plume crossed the Cascades into the Inland Empire. At the same time, an arctic cold front swept across Canada and began to slip into eastern Washington and northern Idaho on November 18. By late November 18, because of the arctic air from the north, temperatures had cooled just enough to support light snow across the Spokane area. By the morning of November 19, 2 to 4 inches of wet snow covered Spokane. Then the core of the pipeline of relatively warm, tropical air, traveling at 8,000 feet, breached the Cascades. At the surface, temperatures hovered between 29 and 31°F. These conditions were perfect for a severe ice storm: moist, warm air overrunning a cold, subfreezing layer of air near the surface.

At 8:00 AM on November 19, gentle snow turned into a driving, freezing rainstorm. The warm, tropical air heated the middle atmosphere (about 8,000 to 12,000 feet) to above freezing, while cold, dense air hovered near the ground, keeping the surface layer of air below freezing. As the middle layer shed rain, it fell through the lower layer and became a continual downpour of heavy, freezing rain. For twelve hours these downpours socked the Spokane area, coating everything in sight with a combination of snow and ice. Between 4:00 PM on November 18 and 4:00 PM on November 19, Spokane International Airport reported 1.10 inches of precipitation—about half the normal total November precipitation.

As the temperature rose, the freezing rain began to taper off, and an afternoon northwesterly wind of 10 to 30 miles per hour picked up and buffeted the city. The strong winds caused conditions to go from bad to extremely dangerous. Ice-laden power lines and trees began snapping.

Cyclists wind through a surreal world of ice along the Centennial Trail in Spokane, Washington, on November 20, 1998. —Dan McComb photo, *The Spokesman-Review*

Roads were mostly wet but not icy due to stored heat in the ground from summer. Surface temperatures rose slightly, and the city carried out de-icing efforts. Nonetheless, downed power lines and fallen trees made roads impassable. By 8:00 PM, officials declared Spokane County a federal disaster area. Then the city of 178,000 people fell into darkness. At one point, one-quarter million people in eastern Washington, including 70 percent of Spokane's residents, had no power. For many, the electricity remained off until December 3, fourteen days later. Washington Water and Power reported that 40 percent of the city's power lines lay frozen to the ground after the storm.

In the aftermath of this extreme storm, Spokane looked like a war zone. Nearly every tree sustained some damage—one-third of the region's trees died. Falling trees severely damaged five hundred homes and businesses. In many cases, the tops of trees snapped and fell like missiles into homes and buildings. In all, the storm killed at least five people and cost at least $22 million.

In Idaho, the storm dropped 2 to 3 feet of heavy snow around Bonners Ferry, collapsing the roofs of several businesses, schools, and homes. Freezing rain coated everything with 1 inch of ice in Kootenai, Clearwater, and Idaho Counties. Damage in Idaho alone reached $6 million. The National Weather Service ranks the event as Idaho's seventh top weather event of the twentieth century.

The November 1996 ice storm stands out as severe and unusual. On average, Spokane receives eight hours of freezing rain a year, compared to seventeen hours for Pendleton, Oregon, and twelve hours for Portland, Oregon, and Yakima, Washington. On the other hand, Spokane receives more hours of freezing drizzle—on average eighteen hours per year—than any other city in the Columbia Basin of eastern Washington and northern Oregon. Drizzle is defined as small, slowly falling water droplets with diameters between 0.20 and 0.50 millimeters; rain is defined as water droplets greater than 0.50 millimeters in diameter. Although freezing drizzle poses less of a threat than freezing rain, it can quickly glaze streets and sidewalks. Icing events typically occur during the morning between 1:00 AM and 9:00 AM, when surface temperatures dip; the afternoon, from 11:00 AM to 4:00 PM, is the least likely time for freezing rain or drizzle.

The usually light, freezing rain/drizzle storms that hit the Columbia Basin leave only 0.10 inch of measurable precipitation per event. As a whole, freezing precipitation throughout the year accounts for 2 to 4 percent of the total annual precipitation across the Columbia Basin, enough to make it the ice-storm belt of the West. In the United States, ice storms wreak the most havoc in the Midwest, particularly in eastern Nebraska, Iowa, and Missouri, where forty or more hours of freezing rain occurs a year. This compares with fifteen to thirty hours in the Spokane-Pendleton region.

4

California's Coast, Western Valleys, and Southern Mountains

Elevation: sea level to 8,800 feet

Natural vegetation: broadleaf deciduous and needleleaf evergreen trees in the north-central part of this region; chaparral in the southern portion; more grasses and other herbaceous plants in interior valleys

Köppen climate classification: Mediterranean, with moist, mild winters, and dry, cool summers; western valleys with long hot summers fall into a different classification

Most dominant climate feature: short wintertime wet season, otherwise dry, pleasant weather

The wet season reaches its peak intensity over much of California in January and February, when frequent Pacific storms spread rain and snow across the state. Average precipitation totals for January and February range from as low as 3 inches in the southern valleys to as high as 20 inches in the northern Coast Ranges. Since the storm track frequently shifts southward across California at this time of year, Los Angeles typically receives nearly 3 inches of rain in January alone. Also in January, average daily temperatures range from the cool 40s in the north to the 50s along the southern coast and warm slightly in February. Hard freezes are most likely to occur at this time of year.

By March and April, the powerful storms that brought heavy rains and high winds earlier in the year become less frequent. The Pacific storm track centered over central and southern California begins to shift northward, and the winter wet season wanes. Precipitation totals for March and April range from nearly 12 inches in the higher terrain of the north to less than 2 inches in the south. Average temperatures warm into the 50s and 60s. The last spring freeze typically occurs during these months, except along the coast, where freezes rarely happen at any time of the year.

Precipitation becomes scarce in May and nonexistent in June. Most areas receive less than an inch in May. Temperatures, particularly inland, turn hot. Daytime readings soar into the 90s, while the coastal and higher elevations reach the 70s and 80s. Morning lows range wildly from the 40s to the 60s depending on elevation and proximity to ocean.

The dry, hot conditions that began in June continue through July and August. Frequent thermal heat lows form over the deserts and migrate northward through the western valleys, pushing temperatures to their highest levels of the year. In summer, average high temperatures hit the 90s and 100s throughout the inland valleys, while cool marine air moderates high temperatures along the coast to the 70s and 80s. Nighttime low temperatures are mainly in the 50s and 60s. Little, if any, precipitation falls during these months.

September and October offer the warmest daytime temperatures of the year along the

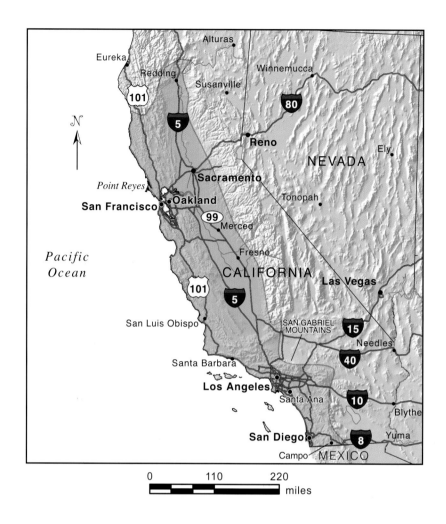

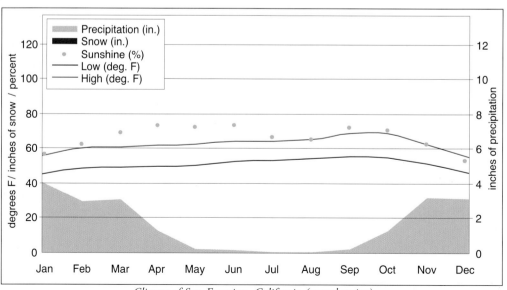

Climate of San Francisco, California (coastal region)

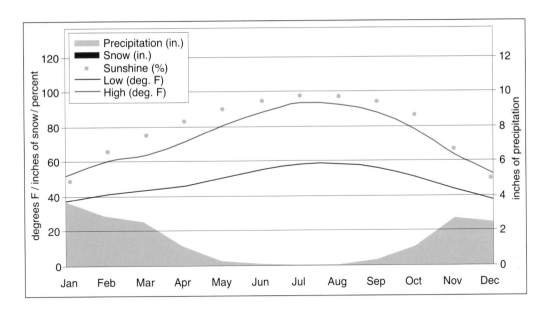

Climate of Sacramento, California (inland valley)

immediate coast with highs in the low 60s, while precipitation is light; however, the first rain of the wet season begins falling in late October. Most areas receive less than 1 inch of precipitation on average in September, but totals range up to 4 inches in the north in October. Average daily temperatures slowly cool into the 60s and 70s by October.

Although the mountainous areas feel the first fall freeze in late October, most inland areas experience the first frost in late November, or not at all along the southern coast. Temperatures during November and December become relatively cool, with daytime highs in the 50s and 60s and nighttime lows mainly in the 40s and 50s. Temperatures and rain decrease in November and December. After a bone-dry summer, as much as 20 inches of precipitation can fall in the higher terrain of the north in November and December, while as little as 1 inch dampens the southern valleys. Some of this moisture falls in the form of snow at the highest elevations, particularly in the mountains east of Los Angeles, where up to 60 inches of this region's total precipitation can fall during these months. The Coast Ranges of northern California can receive up to 24 inches of snow annually, and most of it falls at this time of year.

The Fog Belt: San Francisco

Thick fog delayed the discovery of the San Francisco Bay area for some two hundred years. The English navigator Sir Francis Drake and his crew could not locate the entrance to the fog-shrouded bay in 1579. For nearly a month, they searched for a way in, but the "thick mists and most stinging fogs" left them hopeless. Not until the eighteenth century did the Spaniards finally find their way into the bay. Because of this and the experiences of other travelers and residents, we know the Pacific Coast from Point Conception, California, to northern Washington as the "fog belt" of the West.

When visibility decreases to 3 miles or less because of moisture suspended in the atmosphere, the National Weather Service considers it "fog." Fog along the Pacific Coast most commonly occurs in summer when upwelling—the transport of cold water near the ocean floor to the surface—causes sea surface temperatures along the coastline to drop. The cold water cools the air above it, making the atmosphere stable. The cooled air sinks and behaves like air in a high-pressure system. As the sun heats the ocean surface, water evaporates and quickly condenses amid the cold air and forms fog. Because of the stable atmosphere, the moisture in the air remains as fog and does not fall as rain. At the same time fog forms, warm inland temperatures create relatively low pressure. As the fog moves inland toward the lower pressure it encounters the warm air, causing the fog droplets to convert (evaporate) into vapor.

The northern and central California coasts have some of the foggiest places in the United States. Point Reyes, California, 35 miles north-northwest of San Francisco, averages 1,468 hours of fog each year—17 percent of the year. However, the foggiest location in the United States, Cape Disappointment, Washington, at the mouth of the Columbia River, gets blanketed in fog an average 2,556 hours a year—29 percent of the time.

The World's Foggiest Areas

In the United States the three foggiest regions are the Pacific Coast states, New England, and the higher terrain of the Appalachian Mountains. Globally, the north pole, coastal Antarctica, and the tropical highlands of Central and South America have a significant number of days with fog. Newfoundland is also extremely foggy, with Argentia averaging 206 days of fog a year. In general, these areas have water nearby that evaporates, adding water vapor to the air, which then reacts with cool temperatures to form fog.

This high-resolution (1 kilometer) satellite image shows fog stretching along the coast from San Francisco to Los Angeles on July 7, 1997. —Image courtesy of the National Oceanic and Atmospheric Administration

Thick fog has contributed to some catastrophes at sea. One of the deadliest occurred on February 22, 1901, when the Pacific Mail Fleet's passenger steamer *City of Rio de Janeiro* struck a rock while entering the foggy San Francisco Bay. One hundred twenty-eight people, mostly Asian immigrants, died. Another catastrophe occurred on August 6, 1921, when the Pacific steamer *Alaska*, en route to San Francisco from Seattle, hit Blunts Reef off Cape Mendocino in thick fog and forty-five people died.

Although some dispute it, many consider the Point Honda Disaster of September 8, 1923, fog related. A squadron of U.S. Navy destroyers traveling from San Francisco to San Diego experienced fog and reduced visibility of less than ½ mile. The dark of night combined with the dense fog forced the squadron's commander to

change course and make an approach to Santa Barbara Channel. Minutes later the USS *Delph* ran aground near the jagged rocks off Point Arguello (just west of Lompoc), known locally as Point Honda or Devil's Jaw. The squadron commander immediately sent warning signals

Fog Drip

As heavy moisture-laden fog blows around trees or other objects, it leaves moisture on their surfaces. That moisture, known as "fog drip," then falls to the ground. In the Northwest, locals call fogs that carry a lot of moisture "Oregon mist." Oregon mist forms far out to sea. The cold surface temperature of the Pacific causes relatively warm, moist air to cool and condense into fog. Onshore breezes then carry the fog inland, where it collects on exposed objects, eventually dropping to the ground as a light shower.

Fog drip provides an extremely important source of moisture for coastal vegetation in California and Oregon. Over the course of a summer, the Berkeley Hills can receive up to 10 inches of precipitation from fog drip, nearly half Berkeley's annual rainfall. Likewise, fog drip can amount to as much as 0.59 inch of precipitation per week at higher elevations on the Monterey Peninsula. The heavily forested areas of the northern California coast, such as Redwood National Park, receive as much as 60 inches of fog drip a year. It plays an important role during the dry summer months since 0.05 inch of fog drip can collect during a single, rainless night—the equivalent of a light rain shower. Overall, fog drip can amount to as much as half the water supply of the redwood forests. While the redwoods themselves collect the fog drip, other species of plants benefit from the large quantities of water. Fog drip from redwoods may also replenish streams and groundwater, but scientists need more data to prove this conclusively.

Lost in a Fog

The foggiest place in the eastern United States—the Libby Islands off the coast of Maine—receives an average of 1,554 hours of fog each year. In 1907 the Seguin Lighthouse reported an East Coast record of 2,734 hours of fog.

Fog moving into the San Francisco Bay obscures the north side of the Golden Gate Bridge. —Photo courtesy of the National Oceanic and Atmospheric Administration/Department of Commerce

to the other vessels, but the configuration of the coastline, the outlying rocks, and the ships' tight formation combined to quickly strand six other vessels on a rocky promontory at night. Of the eight hundred individuals in peril, twenty-three died. The seven vessels were declared total wrecks and later sold as hulks.

Fog is a nuisance not only along the coast but also across inland areas of California and Oregon. Although distant from the Pacific Ocean's foggy influence, these areas fall prey to radiational fog. Generally, radiational fog forms over swamps, rivers, bogs, wetlands, and lakes on clear cold nights; however, with the right conditions—saturated soil, light wind, initially clear skies, and high humidity—radiational fog can occur anywhere. Also called "ground fog," radiational fog forms when the temperature of moist air nearest the surface cools to the dew point, which allows the water vapor to condense into water droplets. The native tules and other marsh plants of these ecosystems give rise to the term "tule fog," a regional name for radiational fog. Dense tule fog plays a role in deadly multivehicle accidents each year. Road signs now warn motorists of the potential for fog to engulf highways.

On December 11, 1997, a fiery accident involving eight tractor-trailers and twelve cars occurred along Interstate 5 near Sacramento, California. Caused by patchy and very dense fog—the most deadly type—the accident claimed the lives of five people and injured twenty-eight others. This type of fog kills because its density makes it difficult to see through, and given its patchiness, drivers often encounter it at high rates of speed. By the time they can see that traffic ahead has slowed or stopped for fog, it is too late to avoid a collision.

El Niño Storms

The weather-disrupting phenomenon known as El Niño rose to prominence in the news during the record-breaking 1997–98 episode. El Niño, for many years only discussed by meteorologists and oceanographers, increasingly became a household term understood and feared by the general public. While some Californians may not know exactly what it is, they know the influence it has on their weather.

Scientists named the weather phenomenon El Niño—Spanish for "the little boy" or "the infant Jesus"—because the abnormally warm sea surface temperatures characteristic of this event typically appear off the coast of South America around Christmas. The term El Niño is shorthand for what meteorologists and scientists call the El Niño/Southern Oscillation, or ENSO.

El Niño is a complex reversal of atmospheric and oceanic conditions in the tropical Pacific Ocean that has implications across the globe. Meteorologists think an El Niño episode is triggered when steady westward-blowing trade winds weaken and even reverse direction, but they don't know what causes this to happen. This change in the winds allows a large mass of unusually warm water that normally sits near Australia to move eastward along the equator until it reaches the coast of South America. Furthermore, the sea-surface warming off the South American coast is amplified by weakened upwelling, which is diminished because of the weakened trade winds. This displaced pool of water, mainly near Peru, results in increased evaporation, thus increased water vapor and clouds in the atmosphere—enough moisture to alter the typical atmospheric jet stream patterns around the world.

The changes in the wind strength and direction of the jet stream typically bring much wetter winters to the southern tier of the United States. Sea level tends to rise as well since surface winds "pile up" the Pacific Ocean's waters along California's coast. Strong winds and low pressure associated with low-pressure systems will also cause seas to rise around a storm's center. When this mound of elevated sea level moves ashore, it is called *storm surge*.

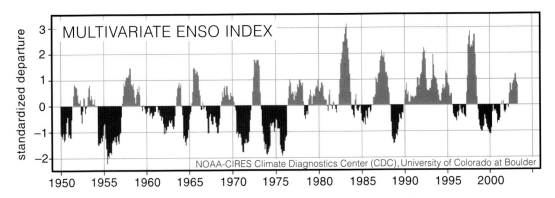

To best monitor the coupled oceanic-atmospheric character of ENSO, meteorologists developed the Multivariate ENSO Index (MEI). One of many indexes used by meteorologists to categorize and study ENSO, the MEI is a weighted average of the main ENSO variables observed over the equatorial Pacific. These include sea level pressure, east-west and north-south components of surface wind, sea surface temperatures, surface air temperature, and total amount of cloudiness. A positive MEI indicates an El Niño episode; a negative value indicates a La Niña event. The peaks represent particularly strong El Niños and La Niñas that contained the strongest ENSO variables. —Courtesy of NOAA-CIRES Climate Diagnostics Center*

The increased amount of available water vapor injected into the subtropical jet stream, a high-altitude river of fast-flowing air, pummels California with a continual stream of storms. More intense storm activity from the Pineapple Express (see *chapter 1*) occurs during El Niño years. A Pineapple Express commonly leads to prolonged periods of extremely heavy rain, mountain snow, and hurricane-force winds along the West Coast. El Niño demonstrates how the ocean and atmosphere work together to dictate weather and climate.

El Niño events, also called "warm events," occur about every two to seven years. The opposite of an El Niño is a La Niña, which translates to "little girl." La Niñas are also sometimes called *El Viejo, anti–El Niño,* or simply a *cold event* or a *cold episode*. La Niñas are characterized by unusually cold ocean temperatures in the equatorial Pacific, which alter the typical atmospheric jet stream patterns around the world.

In the United States, residents see and feel the influence of either an El Niño or La Niña most profoundly during winter. Because the temperature difference between the north pole and equator is greatest during winter, the atmosphere is very energetic; therefore, any disruption in the world's wind pattern by an ENSO event translates and magnifies into stronger storms (low-pressure areas) in some places and larger areas of high pressure (clear skies) in others. In general, El Niño delivers California a wetter and stormier winter while leaving the Pacific Northwest warm and dry. On the other hand, La Niña greets California with much calmer, drier, and warmer winters while slamming the Pacific Northwest with winter storms.

During the major 1982–83 El Niño, a slow-moving Rossby wave propagated up the West Coast and increased sea levels 4 inches in San Francisco. The shape and rotation of the earth govern Rossby waves, also known as *planetary waves*. In the atmosphere, Rossby waves are observed as large-scale meanders of the mid-latitude jet stream. Carl-Gustav Rossby first

Climate division anomalies associated with El Niño (1957–58, 1965–66, 1968–69, 1972–73, 1982–83, 1986–87, 1991–92, 1997–98) and La Niña events (1954–55, 1955–56, 1964–65, 1970–71, 1973–74, 1975–76, 1988–89, 1998–99) based on long-term winter temperature and precipitation averages from 1971 through 2000. For example, the northwestern California climate area will receive 4 more inches of winter precipitation (December through February) than normal during an El Niño–enhanced winter. The patterns of these maps are more important than the magnitude of the anomaly. For instance, during a La Niña winter, precipitation tends to be heavier than normal in the Pacific Northwest. —Courtesy of NOAA-CIRES/Climate Diagnostics Center

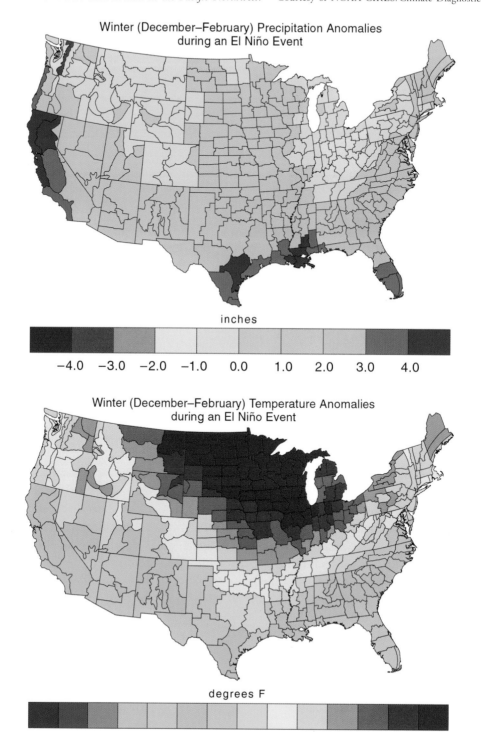

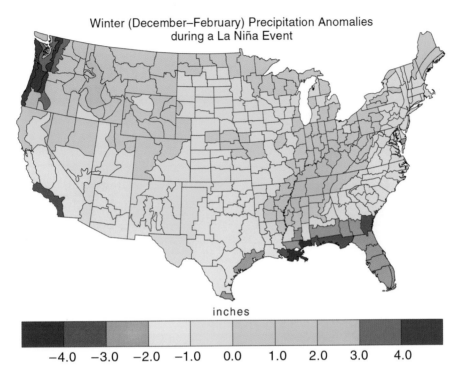

Winter (December–February) Precipitation Anomalies
during a La Niña Event

inches

−4.0 −3.0 −2.0 −1.0 0.0 1.0 2.0 3.0 4.0

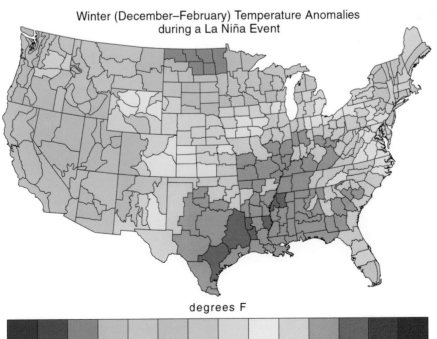

Winter (December–February) Temperature Anomalies
during a La Niña Event

degrees F

−3.0 −2.0 −1.0 0.0 1.0 2.0 3.0

This March 3, 1998, photograph shows a hillside eroded by El Niño–powered storms in Studio City, California. Landslides crushed the homes beneath the ones pictured. —Dave Gatley photo, Federal Emergency Management Agency

El Niño–powered storms caused disastrous mud slides in Rio Nido, west of Santa Rosa, California. —Dave Gatley photo, Federal Emergency Management Agency

theorized about their existence in the oceans in the 1930s, and only recently (with the advent of satellite oceanography) did scientists confirm their existence. The sea level increase, coupled with high astronomical tides, heavy river flows, record rains, and strong storms, made the coast unusually vulnerable to erosion and damage caused by high storm waves. The 1982–83 storms caused an estimated $50 million ($90.3 million in 2002 dollars) in damage to public beach and pier structures in California alone. High waves and flooding damaged or destroyed 6,661 homes and 1,330 businesses, and estimates of damage to private property reached the hundreds of millions of dollars. The event killed 36 and injured 481, and officials estimated total losses for all of California reached $1.2 billion ($2.2 billion in 2002 dollars).

The 1997–98 El Niño, noteworthy but not as damaging as the 1982–83 episode, caused death and destruction in California in January, February, and March 1998. A parade of powerful storms slammed the California Coast with hurricane-force winds, heavy rains, copious mountain snows, and high storm tides. Because the barrage was continual, residents did not have time to mop up after an El Niño–inspired storm before the next one arrived. In California, precipitation totals by the end of February shot past the state's normal annual precipitation. These record-breaking rains soaked California, sending mudslides out of control and rivers over their banks. University of California at Riverside reported 9.49 inches in February—the all-time greatest one-month precipitation total at the university since record keeping began in 1924. Seventeen people died, most from drowning.

The 1982–83 and 1997–98 El Niños left a lasting impression on California residents. And while the words *El Niño* may bring fear to some Californians, the typical El Niño does not result in catastrophe, but simply a wetter than normal winter. The National Oceanic and Atmospheric

Administration considers the 1982–83 and 1997–98 El Niños two of the nation's top weather events of the twentieth century.

Santa Ana, Diablo, and the Catalina Eddy: California Winds

Because of southern California's location between the ocean and inland deserts and its mountainous topography, Californians experience several hot, dry winds of the foehn (pronounced firn) wind family. Foehn winds are warm, dry winds that blow down the sides of mountains. Like the Chinook east of the Rockies, they can bring a pleasant change in weather; sometimes, though, they can turn parts of California into raging infernos.

Santa Ana winds, named after the Santa Ana Mountains and canyons they blow down, are warm, dry, gusty winds that buffet the valleys of southern California. True Santa Ana winds only strike areas south of Los Angeles, while similar winds (also called "Santa Anas") occasionally sweep through areas to the north. The "witch's wind" is just one of many nicknames for the Santa Anas.

These potentially dangerous and deadly winds occur year-round, but most commonly between October and February. December has the highest frequency of Santa Ana events because during this month a high-pressure system is more likely to reside over the Great Basin. Summer Santa Ana winds are rare. Santa Anas typically blow northerly to easterly at 40 miles per hour (wind speeds must exceed 29 miles per hour for forecasters at the National Weather Service in Oxnard and San Diego to consider a wind a "Santa Ana"), with gusts up to 58 miles per hour racing over passes and out the mouths of canyons. Stronger Santa Ana winds can produce gusts greater than 69 miles per hour over widespread areas and gusts greater than 115 miles per hour in favorable areas near canyon mouths. The strongest winds typically blow during the night and morning hours when a sea breeze is absent. The sea breeze, which typically blows onshore, counteracts the Santa Ana winds during the late morning and afternoon hours. Pine Valley, in San Diego County along Interstate 8, commonly suffers the highest Santa Ana wind speeds. During a typical Santa Ana event on December 27, 1997, winds gusted to 96 miles per hour and blew several eighteen-wheelers off the highway in Pine Valley.

The force of a pressure gradient between a high-pressure area (clockwise winds) over the Great Basin, typically near the intersection of Idaho, Utah, and Nevada, and a low-pressure area (counterclockwise winds) off the southern California coast generates these strong winds. Winds wrap around these two atmospheric circulations and shoot down the canyons of the Santa Ana, San Gabriel, Santa Monica, and San Bernardino Mountains north and east of Los Angeles. Furthermore, upper atmosphere turbulence sometimes mixes strong northerly or

easterly winds aloft down to the surface, increasing Santa Ana wind speeds. The air accelerates as it funnels through the narrow canyons, and as it descends about 9,000 feet its temperature in-

Sundowners

Northerly downslope winds on the lee side of the Santa Ynez Mountains, called "sundowners," hug the coast north of Santa Barbara. These winds, like other offshore winds in southern California, are largely due to a pressure gradient between high pressure inland and lower pressure off the coast. Sundowners cause anomalously high temperatures in the Santa Barbara area—over 100°F—and produce severe wind gusts. They occur an average of three to four times a year, mainly blowing between 1:00 and 4:00 PM, and last a few hours, giving rise to their name. On June 27, 1990, a particularly severe sundowner flamed an arson-ignited fire, which destroyed over five hundred structures in Santa Barbara.

The meteorological setting for Santa Ana winds. Clockwise winds around a high-pressure system over the Great Basin cause a northerly to easterly flow over southern California. The counterclockwise circulation and pressure of a weak low offshore strengthens this flow. The air warms and dries as it descends the mountains, causing the Santa Ana's searing effects.

creases through adiabatic warming. Consider a parcel of air at some high elevation. If we move the parcel to a lower elevation, the air pressure decreases with decreasing elevation, forcing the parcel to compress. This compression causes air molecules within the parcel to gain kinetic, or internal, energy and begin to move faster. As the air molecules speed up, the temperature increases

Disastrous Santa Ana Wind Events

November 1961: Weather conditions combined to create a large fire that torched Bel Air. A prolonged drought of many years had reduced the moisture content of vegetation to almost zero. Hot, dry 55- to 65-mile-per-hour Santa Ana winds blew through the region and drove the humidity to very low levels. Unseasonably hot temperatures registered in the 90s and 100s. Once the fire started (its cause unknown), thermal air currents created by the intense heat, and high-velocity winds in advance of the main fire front, tossed burning coals and branches into the air. Bel Air had few streets, and those were narrow and winding. Incoming fire equipment and exiting residents jammed the streets, and firefighters had to walk to the scene and combat the fire with hand tools. Under such adverse conditions, natural and man made fire barriers could not interrupt the fire's progress. Many residents lost all the possessions they left behind. One hundred thirty firemen received injuries; the fire destroyed 484 buildings (mostly residential) and 6,090 acres of wildland, causing $100 million ($602 million in 2002 dollars) in damage. The National Weather Service ranks this as the fourteenth top weather-related event in California in the twentieth century.

September 25, 1970: A Santa Ana pushed the temperature to 105°F at Los Angeles International Airport, just 5°F shy of the site's all-time record high temperature recorded in September 1963. The hot winds dried vegetation, and subsequent fires raged out of control. They hit San Diego the hardest, consuming five hundred homes and five hundred other structures.

September 22–29, 1971: A weeklong Santa Ana fanned several disastrous wildfires from the Oakland/Berkeley area to San Diego. The fires destroyed more than five hundred homes and charred a half million acres.

November 2, 1993: Fierce, hot Santa Ana winds whipped up one of California's most devastating infernos, known as the "Malibu firestorm." Fires charred close to 200,000 acres of hills and canyons in southern California and nearly devastated areas of Santa Monica—one of the country's most exclusive and expensive communities. By month's end, fires had burned two thousand homes, killed four people, and injured twenty-one. Damage estimates reached $1 billion, making this one of the West's few billion-dollar weather catastrophes.

October 26–November 7, 2003: Strong Santa Ana winds pushed thirteen scattered wildfires from the suburbs northwest of Los Angeles all the way to Ensenada, Mexico, about 60 miles south of the border. An extended drought, widespread tree death resulting from insect infestation, and continuing expansion of the urban-wildland interface combined to exacerbate the fires. Twenty-two deaths were blamed on the fires and four thousand homes were destroyed in California. The wildfires, which consumed 750,000 acres (an area 81,000 acres larger than the entire state of Rhode Island), were some of the most destructive and deadly in California's history. The "2003 Firestorm," as the media called it, caused billions of dollars in damage.

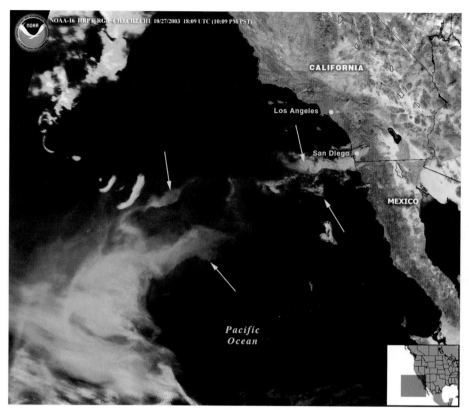

Wildfire smoke (yellow arrows) blows across southern California and the Pacific Ocean in this satellite image taken at 1:09 PM (eastern standard time) on October 27, 2003.
—Image courtesy of Operational Significant Event Imagery support team and the Environmental Applications Team (EAT) of the Satellite Services Division (SSD) of National Environmental Satellite, Data, and Information Service (NESDIS)

within the parcel. Not only do the winds warm but they also dry out. Any moisture they carry evaporates from their increased temperature, drastically increasing the fire danger associated with them.

Santa Ana winds can turn vegetation tinder-dry in a matter of hours as they drive the relative humidity as low as 5 percent and temperatures into the 90s and 100s. Moreover, these winds can reach damaging speeds of 100 miles per hour at canyon mouths and up to 50 miles per hour in the corridor between Los Angeles and San Diego. A blistering Santa Ana can quickly fan fires into firestorms that engulf southern California in smoke, dust, heat, and fierce winds. Nearly every year, southern California experiences a Santa Ana event that generates a disastrous fire.

On a positive note, the Santa Ana winds quickly scour the Los Angeles basin, clearing out any stagnant or smoggy air; however, during the most severe Santa Ana windstorms, smoke and dust replace the smoggy air. The winds can also cause unusually high waves on the northeast side of the Channel Islands, which makes for good surfing.

The Catalina Eddy

The curved coastline and coastal topography of southern California creates a number of unique local wind patterns. One of the most famous is the Catalina Eddy, a weak cyclonic or counterclockwise circulation that forms off the coast of Los Angeles. The summer months—known as the "stratus season" to meteorologists—from April through September provide the most favorable conditions for a Catalina Eddy to form because the strong Pacific storms of winter hinder its somewhat

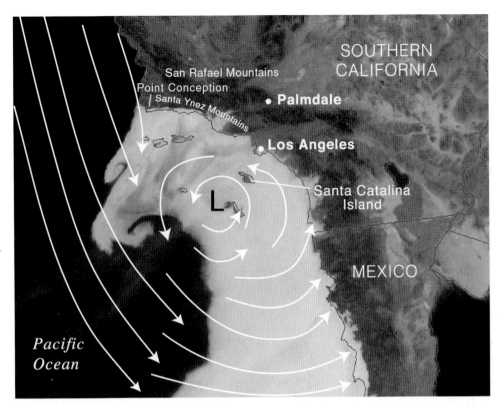

A satellite image of a classic Catalina Eddy off the coast of Southern California, taken on June 6, 2002. When northwesterly winds along the southern California coast blow stronger than normal, they interact with the local coastal topography to form an atmospheric counterclockwise vortex off Los Angeles. This vortex gets its name because it forms around Santa Catalina Island. The gentle winds of the eddy often direct the offshore marine layer toward the Los Angeles basin. The eddy is only about 60 miles in diameter. —Image courtesy of the National Oceanic and Atmospheric Administration (annotated by author)

delicate circulation. The occurrence of Catalina Eddies peaks in June.

An eddy develops when typical changes in the weather lead to a strong east-west pressure difference, or gradient, between the ocean and the mainland. As high pressure builds over the ocean, a low-pressure system, called a *heat low,* spins inland over the southeastern California deserts, resulting in a northwesterly wind. As the heat low intensifies, the pressure gradient strengthens, as does the northwesterly wind along the coast and over the San Rafael and Santa Ynez Mountains east of Point Conception. The winds flowing over the mountain ranges and around Point Conception cause a counterclockwise circulation—known as a *lee eddy*—to form near Catalina Island, and the famous Catalina

Eddy is born. Topographically trapped, the circulation can maintain itself for days.

The southerly winds that develop as part of the vortex along the Los Angeles coast bring in a deep, cool, cloudy layer of air. This weather provides a welcome relief from the hot summertime temperatures and can aid firefighters battling brush fires at this time of year. Sometimes the winds deliver light drizzle and fog in the morning hours along the coast, hindering aviation. A typical Catalina Eddy will allow low coastal clouds and fog to persist into the afternoon; normally they burn off by late morning. In fact, when the Catalina Eddy peaks, the depth of the low clouds may extend to 6,000 feet and push through the inland valleys, reaching as far as Palmdale.

The Diablo, or Mono, winds are hot, dry, northeasterly winds that sweep down the hills east of San Francisco through the canyons of the Diablo Range, which runs north-south on the east side of the Bay Area. Diablo winds are a type of foehn wind, a wind that occurs when a thick layer of prevailing wind is forced over a mountain range. The name *Diablo,* Spanish for "devil," suits these hot, dry winds. Eventually these winds sweep into the Bay Area and cause severe wildfires to grow exponentially.

Diablos can develop during May and October, but are most common in fall when high pressure builds over Nevada or Oregon and lower pressure forms along the central and southern California coast. This pressure system resembles the one that spawns Santa Anas, but the high pressure develops further north for Diablos. The circulation around these large features forces air moving west from the San Joaquin Valley to accelerate as it descends the west, or lee, side of the Berkeley Hills. As the winds descend the mountains, they typically reach 40 to 60 miles per hour and occasionally blast at 70 miles per hour. On November 26, 1976, a 90-mile-per-hour gust was reported.

Like Santa Anas, Diablo winds warm and dry through adiabatic warming as they descend the mountains and can drive temperatures to record highs and humidity to record lows. An increase in air temperature of 20°F causes the relative humidity to drop by about 50 percent. Relative humidity controls the moisture content of fuels such as brush and grass, and thus determines their susceptibility to fire. Fuels with 20 percent moisture can catch fire, but light fuels with 2 percent moisture, such as grasses, dried twigs, and dead underbrush, can explode like gasoline. Drought conditions combined with hot, dry Diablo winds wilt vegetation immediately, turning it to brittle tinder. With the slightest spark, it can burst into an uncontrollable wildfire. These winds often result in disastrous fires, dust storms, and debili-

tating heat waves, thus giving rise to the nickname "hair drier" winds.

The Newhall Wind

Another of Southern California's local foehn winds is known as the "Newhall wind." This wind is caused by a pressure gradient—high pressure to the north, low pressure to the south—which draws air from high on the Mojave Desert to Newhall Pass. The descending wind funnels through Newhall Pass into the San Fernando Valley north of Los Angeles. As the hot desert air moves to lower elevations it warms and dries more from adiabatic warming, bringing an unwelcome furnacelike blast to the Burbank area. Another warm, dry local wind, the Sonora, blows from Sonora, Mexico, into Baja California and southern California in the late spring.

One of the worst firestorms in California—caused largely by a Diablo wind—charred the Oakland and Berkeley Hills east of San Francisco on September 17, 1923. High winds blew catastrophic fires into the northeast section of Berkeley. Within three hours fire had consumed 577 homes. This fire went into the record books as Berkeley's worst natural disaster.

The Oakland firestorm of October 1991—referred to as the Tunnel Fire—ranks as one of the few billion-dollar weather-related events to ever strike the state. The fire originated on a steep hillside above Highway 24 near the entrance to the Caldecott Tunnel. Low humidity and Diablo winds fanned the fire, which caused approximately $1.5 billion ($2 billion in 2002 dollars) in damage and killed twenty-five people. The Federal Emergency Management Agency's Hazard Mitigation Report noted that all the conditions for a major disaster showed up that morning: record high temperatures well into the

An aerial view of a neighborhood burned catastrophically in the 1991 Oakland firestorm, taken on October 25, 1991. Although the fire only burned 1,520 acres, it destroyed 2,843 single-family dwellings. —Photo courtesy of Robert A. Eplett, California Governor's Office of Emergency Services

Smoke from the 1991 Oakland firestorm caused this orange sunset over San Francisco in a photograph taken on October 22, 1991, from the burned neighborhood pictured above. —Photo courtesy of Robert A. Eplett, California Governor's Office of Emergency Services

90s, hot, dry winds, five years of drought-dry brush, and groves of freeze-damaged Monterey pines and eucalyptus trees.

Intense Precipitation: San Gabriel Mountains

Many westerners cannot imagine 26.12 inches of rain in one year, much less imagine that much rain falling in twenty-four hours. But that is what happened between January 22 and 23, 1943, at Hoegees Camp, a weather station perched at an elevation of 2,760 feet on the western side of Mount Wilson (elevation 5,710 feet) in southern California's San Gabriel Mountains. This event stands as California's twenty-four-hour precipitation record, and the record stands for all states west of the hundredth meridian as well. Moreover, it is the sixth highest twenty-four-hour precipitation amount ever recorded in the entire United States.

Top Ten Twenty-Four-Hour Maximum Precipitation Records in the United States

PLACE	DATE	AMOUNT (inches)
Alvin, TX	July 25–26, 1979	43 (estimated)
Yankeetown, FL	Sept. 5, 1950	38.70
Kilauea Plantation, HI	Jan. 24–25, 1956	38
Dauphin Island Sea Lab, AL	July 19–20, 1997	35.52
Smethport, PA	July 17, 1942	34.50
Nelson County, VA	Aug. 20, 1969	27
Hoegees Camp, CA	Jan. 22–23, 1943	26.12
Altapass, NC	July 15–16, 1916	22.22
Hackberry, LA	Aug. 28–29, 1962	22.00

Westerners may find it surprising that the San Gabriel Mountains northeast of Los Angeles—in dry southern California, of all places—holds the highest twenty-four-hour precipitation frequency estimates in the United States. Although other parts of the country have measured phenomenal twenty-four-hour rainfall amounts, the San Gabriel Mountains are statistically more likely to have extremely heavy twenty-four-hour rainfall amounts more often. Meteorologists estimate that the one-hundred-year, twenty-four-hour rainfall total—meaning a 1 percent chance exists that a certain amount of rain will fall in a twenty-four-hour period on any given day in one year—for the upper southeast-facing slopes of Thunder Mountain, about 14 miles northwest of San Bernardino, is 24 inches. In other words, a 1 percent chance exists that 24 inches of rain will fall in twenty-four hours in the San Gabriel Mountains in any given year. For comparison, areas along the Gulf of Mexico prone to hurricane landfalls have twenty-four-hour precipitation frequency estimates on the order of 13 to 15 inches, and there is 0.02 percent chance of a heavy, 24-inch rainstorm happening in a state like Texas in any year.

Intense rainfall soaks southern California primarily during winter, when the jet stream drives copious amounts of moisture into the area from the Pacific Ocean. Unlike the Cascades or even the northern Sierra Nevada in California, the San Gabriel Mountains are closer to very moist, low-level equatorial air that has the capacity to hold more water than the cool air that blows into the Pacific Northwest. The most intense rainfall usually only lasts twenty-four hours or less since storms in this area typically move fast. Although intense storminess usually lasts for a day, a storm can drop intermittent precipitation on the mountains for days. The orographic effects of mountains like Thunder Mountain, at 8,587 feet in elevation in the central San Gabriel Mountains, are largely responsible for the phenomenal rainfall amounts in this region. The upper-level jet stream draws tropical moisture into the area and combines with powerful, very moist low-level southwesterly winds to blow the moisture toward the mountains. As the warm water vapor is forced up the San Gabriel Mountains, the atmospheric pressure on that air decreases rapidly with altitude, causing it to expand and cool (this is an example of adiabatic cooling). As the air cools, the molecules slow down. Because the speed at which the water molecules move determines their phase—that is, whether the molecules are in a gas (fast), liquid, or solid (slow) state—the slowing molecules begin to change from gas to liquid (rain) or solid (hail or snow). Typically, very heavy precipitation is the result. The windward slopes of the San Gabriel Mountains wring out the storms so efficiently that areas on the eastern side of the mountains, such as Death Valley, only receive 2 to 5 inches of rain per year. In comparison, average annual precipitation ranges from 30 inches to locally 50 inches in the San Gabriel Mountains.

Although the San Gabriel Mountains claim the highest one-hundred-year, twenty-four-hour precipitation frequency estimates in the country, they don't hold the record for shorter duration (three-hour) rainstorms. In the southern Great Plains, shorter-duration storms in strong thunderstorm complexes produce higher precipitation frequency estimates. The same holds true along the coast of the Gulf of Mexico, where hurricanes make landfall. Both of these areas have one-hundred-year, three-hour rainfall estimates of 5 to 8 inches, which means a 1 percent chance exists that 5 to 8 inches of precipitation will fall

in a three-hour period on any given day in one year, compared to 2 to 4 inches in the San Gabriel Mountains. These precipitation statistics come from the National Oceanic and Atmospheric Administration's National Weather Service and serve as de facto national standards for rainfall intensity at specified frequencies and durations. Civil engineers use such statistics in the design of a wide range of structures, including urban storm-water drainage and storage systems, dams, and spillways. More recently, however, their use has included a broad array of environmental management and analysis concerns, including the Environmental Protection Agency's pollution discharge regulations and the Federal Emergency Management Administration's floodplain mapping program.

Intense Rainfall Records

The U.S. Eighty-Minute Rainfall Record:
On August 8, 1891, a cloudburst over Campo, California, about 40 miles southeast of San Diego, dropped $11\frac{1}{2}$ inches of rain in just eighty minutes.

The World's Five-Minute Rainfall Record:
On February 5, 1976, Haynes Camp, California, received $2\frac{1}{2}$ inches of rain in just five minutes.

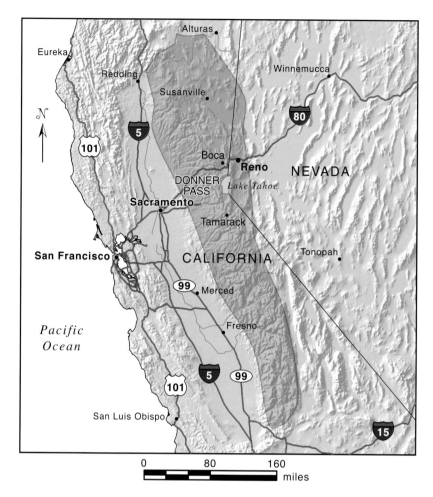

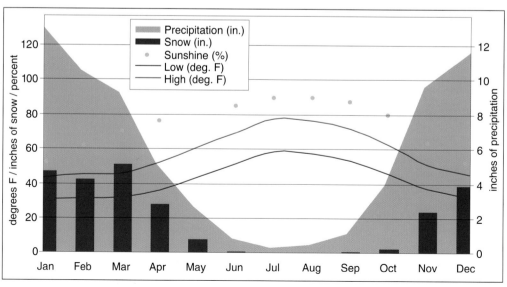

Climate of Blue Canyon, California (percent of total possible sunshine estimated)

5

The Sierra Nevada

Elevation: 2,000 to 14,000 feet

Natural vegetation: needleleaf evergreen trees

Köppen climate classification: Highland

Most dominant climate feature: periods of intense rain and snow during winter

The wet season in the Sierra Nevada is well underway by January and February, and the snowpack grows deep during these months. Up to 24 inches of precipitation—rain and snow—accumulates. The greatest amounts fall when Pacific storms slam into the mountain range and dump their heavy loads of moisture. Temperatures drop to their coolest levels of the year, with average daily temperatures in the 20s and 30s.

Heavy snow and sometimes rain, even at the highest elevations, continue to keep the Sierra Nevada wet through March (up to 12 inches of precipitation), but conditions turn much drier in April (up to 6 inches of precipitation). Average daily temperatures moderate slightly and typically warm to near 40°F by late April. Although most of the highest mountains can experience subfreezing temperatures throughout the year, the lower elevations feel their last spring freeze in late April. Most of the Sierra Nevada's thunderstorms come in April.

The Sierra Nevada is one of the only areas in California that receives significant precipitation in May and June. Most of the range's 1 to 3 inches of precipitation during this period falls in May thunderstorms that the mountains themselves induce through orographic lifting; June is essentially dry. Daytime temperatures rapidly warm into the 60s and 70s by June, but overnight lows still plummet into the 30s. The cold nighttime temperatures are mainly caused by clear skies, dry air, and high elevation, all of which aid efficient radiational cooling.

Summer is in full swing by July, and August brings the region's peak heating and dryness. Precipitation, if any, falls from thunderstorms that develop in stray air masses, particularly in the southern Sierra Nevada. These air masses are not associated with a front or other large-scale atmospheric conditions, and are fed by monsoonal moisture (see *Southwest Monsoon* in *chapter 6*). Daytime temperatures hold at comfortable levels in the 60s and 70s, but again mornings can be chilly, with readings in the 40s. Skies are almost guaranteed to be clear.

September and October are tranquil months with temperatures cooling slowly and precipitation increasing. Before the monsoon wind pattern begins to wane, thunderstorms are still possible in early September; otherwise, precipitation is generated by fronts sweeping down from the north in October. Precipitation is relatively light in September (up to 3 inches in the northern Sierra), while October turns much wetter (up to 10 inches in places). Daytime temperatures begin falling but still remain comfortable in the 50s and 60s; morning lows are typically in the 30s and 40s. Although the first autumn freeze occurs at the very highest elevations in late August, for most areas it occurs in October.

By November and December, the wet season is in full stride. Precipitation totals for these two months range from 10 to 20 inches, mainly falling as snow. Daytime highs are usually in the 40s, while nighttime lows range in the 20s to 30s.

Dangerous Snowstorms

In Spanish, *Sierra Nevada* translates to "snow-covered mountains," and this aptly named range has some of the snowiest places in the world. Powerful and saturated Pacific storms slam into these high-reaching mountains, producing heavy snowfall.

Impressive snowstorms are possible in the Sierra Nevada because of a constant inflow of moisture feeding the growing ice crystals within storm clouds. As with the Cascades, the jet stream collects water vapor over a long fetch of the relatively warm Pacific Ocean, the main moisture source for snowfall in the western United States, and sends it toward the Sierras. Snow is enhanced by orographic lifting: the Sierra Nevada force moisture-laden air to rise into subfreezing temperatures where the water vapor cools and condenses, falling as snow. Snowflakes can consist of single ice crystals or large, multicrystal aggregates. In the Sierra Nevada, the relatively warm (0 to 10°F) inflow of air aloft (14,000 feet) contains more moisture than an inflow of cooler air would. These temperatures and supersaturated conditions in the clouds allow large, six-pointed crystals, known as dendrites, to grow. Because of their size, conglomerates of dendrite snowflakes result in heavy snowfall rates, which explains the extremely high snowfall amounts in the Sierra Nevada.

These heavy snows of the Sierra Nevada can kill. Perhaps the most famous weather-related disaster in the Sierra Nevada, the tragedy of the Donner Party, took place in the winter of 1846–47. The season brought early, heavy, and repeated snowstorms to the Sierra Nevada. The eighty-three-member, California-bound Donner Party arrived in Truckee Meadows—present-day Reno, Nevada—on October 20, 1846, and to their surprise, the mountains ahead were already white with the season's first snowfall.

October snows are not unusual in the Sierra Nevada, but the coincidence of two storms heavy enough to impede travel is rare this early in the season. Emigrants who had crossed the same pass in late November 1844, and others toward the end of December in 1845, had experienced little trouble from the weather.

The Donner Party split up at Truckee Meadows: one group left first and made several attempts to cross the pass (now called Donner Pass), but had to return to Truckee Lake; the other party, slowed by snow even at lower elevations, had to camp at Alder Creek, 5 miles east of Truckee Lake. Earlier emigrants had set up temporary quarters there, and the twenty-one members of the second party took shelter in hastily constructed tents, quilts, and buffalo robes.

The two October storms had covered the pass with about 8 feet of snow. The leading Donner Party made two more attempts to cross the pass during November, but both times deep snow and blizzards turned them back to Truckee Lake. The stranded emigrants were forced to slaughter their remaining cattle and mules for food. Two more storms in early December increased the snow depth at the summit to 12 feet, with 8 feet at Truckee Lake. With only a few intervening periods of fair weather, both parties struggled to survive. The successive storms wore the emigrants down physically and mentally, but the brief interludes of warm sunshine and lulls between the storms gave rise to false hopes that the storms were waning.

In mid-December, a break in storm activity encouraged fifteen party members to make another attempt at the pass on homemade snow-shoes. After the first couple of days, several turned back because of snow blindness, fatigue, frostbite, and starvation. The emigrants struggled to survive at the lake, and with nothing left to eat, they resorted to cannibalism. It took those who had continued on thirty-three days to reach the first ranch on the Sierra's western slope. Tragically, only seven of the people who remained at the lake, including all five women who set out with the original party, survived fatigue and starvation. Meanwhile, back at the Alder Creek camp, emigrants continued to struggle for their lives, and they too resorted to cannibalism.

A series of storms, starting with a Christmas Day blizzard, finally ended on February 6, but not before deepening the snowpack at Truckee Lake to 13 feet. An extended period of calm weather followed this stormy onslaught. During this interlude, two rescue parties from California succeeded in crossing the summit and reached both encampments, but stormy weather forced them to stay at the camps. The last of the survivors who could travel left Truckee Lake on April 21 and made a fast two-day crossing to safety. Between October 16, 1846, and April 3, 1847, ten major storms pummeled the two encampments of the Donner Party, depositing a tremendous amount of snow.

Patrick Breen, a survivor, kept a diary of the Donner Party's nightmare from November 20, 1846, until March 1, 1847, when he stopped accounting the event. In it he documented the weather and snow conditions at Truckee Lake. He mentioned that snow depths fluctuated during winter, but the peak snow depth was "probably about 15 feet to 20 feet at Truckee Lake." Stumps of trees the emigrants cut ranged from 15 to 18 feet tall once the snow melted, indicative of the exceptional snowpack. In all, thirty-eight people lost their lives; forty-five survived. Travelers through the Sierra are reminded of the frightening ordeal of the Donner Party by the pass named in their memory.

California's Coldest Spot

From the blisteringly hot Mojave Desert to the cold, alpine peaks of the Sierra Nevada, the temperature extremes across California boggle the mind. On the morning of January 20, 1937, the temperature fell to −45°F at Boca Reservoir, perched at an elevation of 5,532 feet just 4 miles northeast of Truckee, California. This reading established the state's all-time record low temperature. The January 1937 cold snap hit with a one-two punch. First, a brutally cold air mass swept into the Great Basin during the first week of January 1937, dropping temperatures. Then ten days later a second arctic intrusion spread into the region and reinforced the already entrenched frigid air. Like many of the coldest spots in the West, Boca Reservoir lies in a valley, so cold air from the surrounding mountains settles in it.

Boca Reservoir nearly tied its record low temperature again on February 7, 1989, when the temperature fell to −43°F after a major winter storm finished dumping snow on all of western Nevada and parts of the Sierra Nevada. Heavy snow began falling late in the evening February 1, 1989, and accumulated to depths of 3 feet in the mountains and more than 14 inches in the valleys. As the storm left the area, it pulled an arctic air mass southward behind it, sending temperatures well below zero. In western Nevada, the lows dropped to −31°F in Smith Valley on February 6 and −29°F in Spanish Springs on February 7. The cold weather burst water pipes and caused other property damage. The National Weather Service ranks the event as Nevada's ninth most significant weather event of the twentieth century.

The month of January 1911 was one of the snowiest months ever in the Sierra Nevada. A seemingly endless string of Pacific storms dumped phenomenal amounts of snow on Tamarack, California, located at 6,960 feet near the intersection of Calaveras, Tuolumne, and Alpine Counties in the central Sierra Nevada. Tamarack received 32.5 feet of snow that month, which is still the world record for the greatest single monthly snowfall amount. Furthermore, by the end of March, Tamarack had established the national record snow depth of 37.8 feet—nearly five stories of snow!

The snowiest areas of the Sierra Nevada receive upwards of 250 to 500 inches of snow a year. Blue Canyon, California, ranks as the snowiest location (with reliable weather records) in the lower forty-eight. It receives 254 inches—that's 21.2 feet—of snow each year. Only three places in the United States—Mount Rainier and Mount Baker, Washington, and Crater Lake,

Major Floods in Western Nevada

Because of its proximity to the Sierra Nevada and the Sierra's deep snowpacks, the western Great Basin suffers severe flooding from snowmelt and rain-on-snow events. In midwinter (November through February), Pacific storms can get cut off from the jet steam and spin off the coast of southern California. Meteorologists call these storms *cutoffs*. Since these storms are cut off from the cold air north of the jet stream, they tend to draw in a great deal of warm, moist air from over the Pacific Ocean. Eventually, as a cutoff moves inland, its counterclockwise flow drives widespread rain and higher elevation snow north and northwest into Nevada and California. The combination of heavy rain and melting snow can result in flooding. At other times, persistent and strong Pineapple Expresses provide enough moisture not only for the mountains but also the immediate eastern slopes—this is known as *blow-over precipitation.*

Cleared tracks on the Southern Pacific Railway in 1917 not far from Emigrant Gap near Blue Canyon, California. The Sierra Nevada receives some of the heaviest snowfalls in North America. —Photo courtesy of the National Oceanic and Atmospheric Administration/Department of Commerce

Walker River floodwaters rage through the community of Walker, California—about 70 miles south of Reno, Nevada—in Mono County on January 3, 1997. After receiving upwards of 8 feet of snow in the Sierra Nevada, heavy rains combined with warm temperatures rapidly melted the snow and pushed rivers over their banks.
—Photo courtesy of the Mono County (California) Sheriff's Department

The Flood of 1950: A series of very warm storms moved across western Nevada from November 13 through December 8, melting much of the Sierra Nevada's early snowpack. Subsequent flooding inundated much of downtown Reno, including the Riverside Hotel, which had 4 feet of water on the main floor. The peak discharge of the Truckee River in Reno was 19,900 cubic feet per second on November 21; 419 cubic feet per second is average for November. The East Fork of the Carson River and the West Walker River also flooded. Damage in the Reno area reached $4.4 million ($32.8 million in 2002 dollars), more than two hundred people evacuated their homes, and two people died. The National Weather Service office in Reno ranked this flood as Nevada's third most significant weather event of the twentieth century.

The Flood of 1955: Beginning on December 21, 1955, a large and unseasonably warm storm dropped 10 to 13 inches of precipitation in the Sierra Nevada over a three-day period. Most of this fell as rain

that melted a good deal of the existing snowpack, resulting in major flooding across western Nevada. The overwhelmed Truckee River flooded the Reno airport to a depth of 4 feet. Officials estimated the damage in the Walker, Carson, and Truckee River basins at nearly $4 million dollars ($26.8 million in 2002 dollars). The flood caused numerous injuries and one death.

The New Year's Flood of 1996–97: Intense precipitation and warming caused by a Pineapple Express (see *chapter 1*) dealt Reno a heavy blow, causing an estimated $170 million in damage. Casinos and businesses in the Reno area experienced the brunt of the damage. The Truckee River, which flows east through Reno, reached its highest level on record: 14.97 feet on January 2, 1997. The Reno-Tahoe International Airport closed for more than a day as water lapped at the fuselages of some commercial airliners. The New Year's Flood caused roughly $500 million in damage region wide and is the most damaging and costly flood on record in Nevada.

Oregon—approach the astounding snowfall and snow depth of the Sierra Nevada. And, other cities in the western United States renowned for snow pale in comparison to Blue Canyon's statistics. Denver, Colorado, and Salt Lake City, Utah, receive an average annual snowfall of 63 inches. Locations that receive 500 inches of snow a year occur at higher elevations and have less reliable data.

The Washoe Zephyr: Nevada

In the family of foehn winds, the famous Washoe Zephyr is a moderate westerly wind common in summer along the eastern flank of the Sierra Nevada in western Nevada. The name comes from the area the wind affects, specifically Washoe County, which extends from near Carson City north to the Oregon-California state line. The word *zephyr* is derived from the Greek *zephyros,* which means "west wind."

Mark Twain was among the first to refer to the wind as the Washoe Zephyr. Author and adventurer Twain visited Virginia City in 1892. During his visit, he reported a Washoe Zephyr both "robbed and attacked him." Twain commented that those winds explained why there were so many bald people in Virginia City. He also noted that the winds were of such proportion that they stirred up enough dust to obliterate Nevada Territory from sight.

Like any wind, the Washoe Zephyr is caused by a difference in pressure across an area, known as a pressure gradient. The pressure gradient of the Washoe Zephyr develops between the elevated terrain of the Great Basin, which warms during the day creating low pressure, and areas of central California to the west influenced more by cool marine air, which creates higher pressure. As the pressure gradient increases, so does the strength of the wind. Since the pressure gradient is caused by a temperature difference, the Washoe Zephyr is considered a thermally induced wind, and it can blow on any typical summer afternoon.

The Washoe Zephyr can create severe dust storms across western Nevada. Since Reno, Nevada, and areas around it receive only 7 inches of precipitation a year, there is plenty of dry, sandy soil available to be blown into a turbulent dust cloud.

A zephyr usually lasts three to nine hours and typically blows during the afternoon and evening hours when the pressure gradient is strongest. At Washoe Lake, between Reno and Carson City, the zephyr is predictable and reliable. Each summer day, beginning around 4 PM, the wind speeds increase to 15 to 20 miles per hour then die down when the sun sets. The reliable stiff winds make this and other area lakes popular windsurfing destinations.

Besides Washoe Zephyrs, the eastern slopes of the Sierra Nevada are also prone to typical downslope (katabatic) windstorms associated with strong pressure gradients caused by passing storms and building high-pressure areas. The fiercest katabatic windstorms sweep through during winter and spring when the jet stream races across northern California. Winds are further enhanced when a pressure gradient develops between low pressure (a storm) over the Pacific Northwest and a high-pressure area over southern California—a typical winter weather pattern in the West. When the atmosphere becomes unstable, the jet stream winds aloft mix down to the surface, increasing surface wind speeds.

A particularly strong katabatic windstorm hit areas around Reno, Nevada, on March 12, 1968. The blast downed trees and power lines. Officials reported the strongest wind gust ever at Reno Cannon International Airport at 11:24 PM on January 14, 1988, when a 90-mile-per-hour gust rattled the city. Wind gusts associated with katabatic windstorms in western Nevada can reach devastating speeds of 80 to 110 miles per hour, only slightly less than the famous Chinooks that buffet Boulder, Colorado, and the Front Range.

6
Desert Southwest

Elevation: −300 to 12,000 feet

Natural vegetation: desert shrub and semidesert vegetation

Köppen climate classification: dry desert climate

Most dominant climate features: generally hot and dry, with late summer thunderstorms

In the United States, only southern Texas and Florida are warmer in January and February than the deserts of the Southwest. Although nighttime lows typically dip into the 30s and 40s, daytime highs reach the 60s and 70s. Most precipitation stays in the higher elevations (generally above 7,000 feet), where it falls as rain or snow. Some localized storms can deposit as much as 3 inches of precipitation. The lower elevations are lucky to see even ½ inch in January and February combined.

By late March and April, the Southwest experiences a marked decrease in precipitation as the storm track shifts north and the sun warms and dries the air. Vegetation in the desert can bloom prolifically at this time of year, but the hot temperatures and merciless dry, sunny conditions soon force the vegetation into summer hibernation. Daytime temperatures rapidly climb into the 80s by April, with overnight lows staying in the 40s and 50s through March and April. Precipitation is scarce—only light amounts fall in higher terrain.

By May temperatures over 100°F become common, and this is the rule in June. The hot temperatures heat the air, which rises off the desert's surface and produces the summer's first semipermanent feature, the thermal low. Unlike the moisture-rich low-pressure systems that move across the Pacific Ocean and produce clouds and precipitation, the thermal low lacks moisture and instead brings clear skies, high temperatures, and low humidity; even a single day with rain is rare.

Not until late July or August does rain again become possible with the onset of the Southwest monsoon. The monsoon begins when tropical moisture from the Gulf of Mexico and the Pacific Ocean off Mexico's west coast converge over the region. This results in frequent and sometimes widespread thunderstorms that produce some of the most spectacular lightning displays in the United States. Average monthly precipitation increases tenfold in places like Phoenix and Tucson. July and August are the wettest months. Storms deliver 1 to 8 inches (in localized storms) of precipitation. Despite the increased rainfall, temperatures manage to rise to their highest readings of the year. Daytime highs almost always exceed 100°F in the lowlands and climb into the 80s in the mountains. By August, average temperatures drop off a few degrees from their annual maximum as the monsoons bring more clouds and reduce the amount of sunshine.

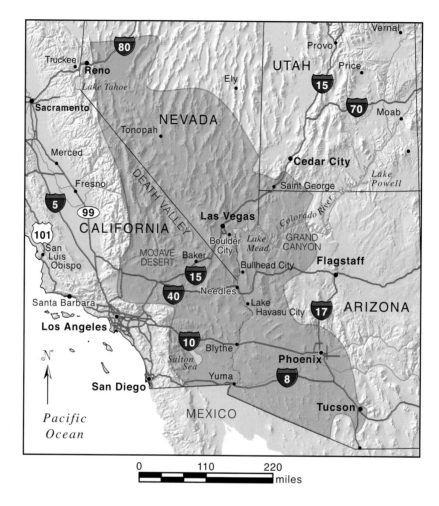

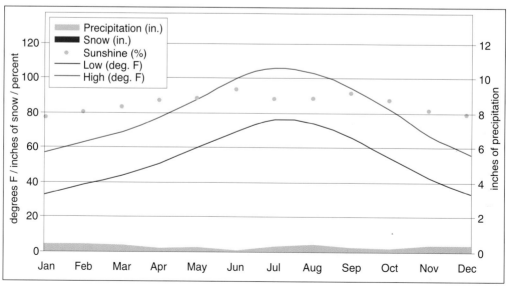

Climate of Las Vegas, Nevada

The summer monsoon begins to wane in September, reducing the average amount of rainfall and increasing the amount of sunshine. Usually by late September, the wet season ends. Temperatures can still be oppressively hot, but by late September the temperatures begin to moderate to more comfortable levels. Strong radiational cooling allows for large diurnal temperature swings, with relatively warm daytime highs and chilly overnight lows. Nighttime lows bottom out in the 50s and 60s, except in the mountains where they reach the 30s and 40s. In the lowlands, daytime readings warm into the 90s and 100s in September then cool into the 80s and 90s by October; the higher elevations see readings about 20°F cooler.

The mild and relatively dry winter of the desert Southwest begins in November and continues through December. Temperatures slowly drop to their coolest readings of the year. Average daily temperatures climb into the 50s and 60s in the lower elevations, and reach the 30s and 40s in the mountains. South-moving Pacific storms can bring episodes of light valley rain and mountain snow, particularly in the higher terrain. On very rare occasions snow can fall even in the lower elevations. In between the occasional storms, sunny and pleasant weather makes the Southwest a popular winter travel destination.

The Nation's Hottest Spots: Death Valley and Yuma

Death Valley is a hot and dusty trough in the Mojave Desert in southern California. Sandwiched between the Sierra Nevada and the Nevada border, this region offers the greatest topographical relief in the contiguous United States: 14,766 feet of difference between Mount Whitney (14,494 feet) in the Sierra Nevada and Badwater Basin (282 feet below sea level) in Death Valley. While the views into Death Valley can be dizzying with steep canyons, sun-baked flats, and ephemeral mirages, the unrelenting heat is scorching. The Mojave

Desert and the Sonoran Desert, which is south of the Mojave, endure the hottest temperatures in North America, particularly during the summer. The extreme national daily high temperature is recorded each year somewhere in southern Nevada, southeastern California, or western Arizona.

Surprisingly enough, the sun does not directly heat the air; longwave radiation from Earth's surface heats the air. In most places, some of the sun's energy evaporates water from vegetation in a process called *transpiration;* however, in the sparsely vegetated desert, nearly 100 percent of the sun's energy goes to directly heating the surface of the ground, which then heats and dries the air with longwave radiation. Likewise, the dry desert air lacks atmospheric moisture to absorb or deflect the sun's rays, allowing for more intense heating. The superheated air—less dense than the cooler air surrounding it—rises off the surface. This induces a low-pressure system, and eventually a deep layer of the lower atmosphere warms. The development of the warm, dry low pressure causes an equal but opposite reaction: a dome of high pressure forms above the low. As the warm air rises into the atmosphere and cools, it eventually reaches the tropopause—the lid of the atmosphere—and is forced to spread outward since it cannot go through the tropopause. This spreading out of air, characteristic of a high-pressure system, perpetuates the heating process by steering storms away from the deserts and allowing for continued clear skies and dry air.

The global wind pattern also contributes to the intense heat in this region. This area of the country is near 30 degrees north latitude, the position of the subtropical ridge, which is a band of high atmospheric pressure formed by several subtropical highs (see *Introduction* for more information about global wind patterns). A similar band exists at 30 degrees south latitude. For these reasons, the deserts of the Southwest experience the United States' most extreme heat.

At 135 feet tall, the world's tallest thermometer at Baker, California, in the Mojave Desert, commemorates the area's extremely high temperatures. —Tom Schweich photo (www.schweich.com)

One of the mysterious moving rocks of the Racetrack, the world-famous playa of northwestern Death Valley. Nobody has ever seen the rocks actually move, but many people believe strong winds are the culprit. Winds of 70 miles per hour can rip through the area; however, in order for the rocks to move, the dry lakebed must be wet, and it rarely rains in Death Valley. —Jon Sullivan photo (www.pdphoto.org)

In the deserts of the Southwest, daytime summer temperatures can easily soar over 110°F, and overnight lows sometimes drop only into the 90s. The intense heat and sunshine can heat the sandy surface to 160°F or more during the peak heating of the afternoon. At Death Valley, the searing heat pushes the daytime temperature to 100°F or above on an average of 141 days a year. Statistically, the hottest day at Death Valley is August 2, when the average temperature is 103°F, the average high is 117°F, and the average low is 89°F. For comparison, the coldest day, January 2, usually brings a low temperature near 35°F and a comfortable high of 60°F.

The hottest temperature ever recorded on the North American continent hit California's Death Valley on July 10, 1913, when the temperature

A Low of 100°F

Imagine an evening forecast of "Mostly clear and hot tonight, with a low temperature near 100°F." It would be difficult to believe, but this forecast occasionally suits Death Valley—the only place in the western hemisphere where the daily minimum temperature sometimes doesn't drop below 100°F.

During the seventy-two years between 1931 and 2003, there were twenty days in which the overnight low temperature was 100°F or hotter at Death Valley. That means on average, the overnight low temperature will not drop below 100°F once every four years, or 0.3 times a year. In Death Valley, the longest string of consecutive days with low temperatures at or above 100°F was three days, July 17–19, 1959, recorded at the Greenland Ranch weather station. Just recently, in 2003, the Death Valley weather station reported two consecutive days with readings of nighttime lows over 100°F: July 23, 102°F, and July 24, 103°F. The 103°F reading tied the hottest nighttime low ever, which was first recorded on July 5, 1970.

COOP

Official air temperatures are carefully measured using either a sensor inside of a shield that protects it from solar radiation or a thermometer inside a white weather-instrument shelter—also known as a *cotton region weather shelter*—positioned 5 feet above the surface. The shield and shelter protect their sensor or thermometer from exposure to direct sunshine, precipitation, and condensation, while at the same time providing the instruments adequate ventilation. A growing number of official locations measure temperatures as well as precipitation. However, volunteers have taken most of the reliable and official readings over the years, and these folks have kept the longest records as well.

The National Weather Service Cooperative Observer Program (COOP) is the nation's surface-weather and climate observing network. More than eleven thousand volunteers take observations on farms, in urban and suburban areas, and at national parks, seashores, and mountaintops. Created in 1890 under the Organic Act, COOP has provided the backbone of climate research ever since. COOP equips observers with thermometers, a weather shelter, and an official rain gauge. To be consistent, the National Weather Service offices around the country use the same equipment for daily temperature and precipitation readings.

In the 1990s the National Weather Service, the Federal Aviation Administration, and the Department of Defense started the Automated Surface Observing Systems (ASOS) program. ASOS serves as the nation's primary surface-weather observing network. ASOS is designed to support weather forecast activities and aviation operations, and, at the same time, support the needs of the meteorological, hydrological, and climatological research communities. Currently the country has over 2,200 ASOS stations, and the number continues to grow.

rose to 134°F at Greenland Ranch. This reading, though disputed, is just 2°F shy of the world's hottest temperature of 136°F set at Al-'Aziziyah, Libya, on September 13, 1922. Al-'Aziziyah sits in the northern Sahara Desert, and in the thirty years of record keeping at Al-'Aziziyah, temperatures have also registered 133°F and 127°F. Death Valley's reading of 134°F takes second place for the world record.

The 134°F reading at Greenland Ranch came during five consecutive days of high temperatures ranging from 129 to 134°F. Many meteorologists consider the reading of 134°F questionable because the instruments were exposed to the elements and also because the maximum temperatures at surrounding stations during the same period did not support the extremely high readings. High readings at other Death Valley locations, however, lead some meteorologists to believe in the authenticity of the 134°F reading. Furnace Creek in Death Valley attained trustworthy readings of 128°F in July 1972 and in June 1994. Furthermore, in June 1994, a park ranger measured 131°F at Badwater in Death Valley with a sling psychrometer, a handheld instrument used to measure the water vapor content of the air. (Badwater is typically a few degrees hotter than Furnace Creek on summer afternoons.) Although the hand-measured reading of 131°F is perhaps more accurate than the 134°F reading of 1913, it cannot enter the climatological record since the ranger measured it with an unofficial instrument.

In July 1998, an unusually large and sprawling upper-level high-pressure system, centered near Mercury, Nevada, helped raise temperatures to daily record highs once again in Death Valley. Upper-level highs rarely get much stronger than the July 1998 event, which produced air pressures of 500 millibars at about 19,685 feet. The average height of the 500-millibar pressure layer above southern California in July is 19,291 feet; in January it is 18,503 feet. Although subtle, these changes in height matter because they are caused by the expansion and contraction of air and indicate the strength of a pressure system. The strong high pressure also came during the climatologically hottest time of year for Death Valley.

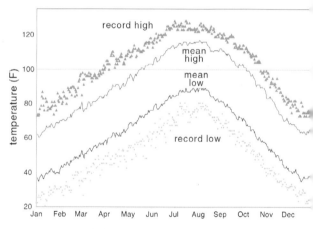

Mean daily maximum, minimum, and extreme temperatures at Death Valley, California

Record High Temperatures for Each Continent

CONTINENT	LOCATION	TEMPERATURE	DATE
Africa	Al-'Aziziyah, Libya	135.9°F	Sept. 13, 1922
Antarctica	Hope Bay	58.3°F	Jan. 5, 1974
Asia	Tirat Tsvi, Israel	129°F	June 21, 1942
Australia	Oodnadatta, South Australia	123.3°F	Jan. 2, 1960
Europe	Seville, Spain	122°F	Aug. 4, 1881
North America	Death Valley, United States	134°F	July 10, 1913
South America	Rivadavia, Argentina	120°F	Dec. 10, 1905

Source: National Climatic Data Center

The heat peaked on July 17, 1998, when maximum temperatures of 129°F and 128°F the next day were recorded at Furnace Creek in Death Valley. Skeptics of the 134°F record proclaimed the 129°F reading was the hottest authentic temperature ever measured in the United States using standard instrumentation at a permanent site. Even though its authenticity is contested, the National Climatic Data Center accepts the 134°F reading as the all-time record high for the United States and the western hemisphere.

Although scientists debate the exact all-time maximum temperature at one or several of Death Valley's monitoring sites, they widely agree that this valley is the hottest spot in North America and the Mojave is one of the hottest deserts in the world. Based on the 134°F event, statistics indicate that Death Valley should expect the temperature to reach 128°F every ten years, 130°F every twenty years, and 131°F every sixty years.

On May 23, 1911, the U.S. Weather Bureau (now called the National Weather Service) in Washington, D.C., named Yuma, Arizona, "America's Hottest City." The story was telegraphed around the country, and when it reached Yuma, it outraged residents. Trying to prove the Weather Bureau wrong, Yuma's offended dug into the weather records for the true answer. They found

hotter places in the United States, namely Death Valley and Lake Havasu, Arizona, but Yuma remains the hottest urban area of the Southwest.

Yuma, the gateway to the Southwest, is not only the hottest city in the United States but also the sunniest and, with 3.01 inches of annual precipitation, the driest. Located in the small Yuma Desert south of the Gila River and east of the Colorado River in the extreme southwestern corner of Arizona, and in the northwestern corner of Sonora, Mexico, Yuma sits at an elevation of 138 feet. The town is an oasis of trees and homes near the California, Arizona, and Mexico border. The sun-loving residents take Yuma's often hostile climate in stride, using gloves to open hot car doors during summer and appropriately naming their newspaper the *Yuma Daily Sun*.

The sun's rays beam down on Yuma an average 4,133 hours each year. Of the total possible sunshine Yuma could see, it receives an amazing 90 percent. On average, the city has only 13 days a year with measurable precipitation. The greatest chance for rain comes in August and September with the onset of monsoon-fed thunderstorms and in February and March when the storm track occasionally dips south into Arizona.

Most of Yuma's precipitation falls in small quantities from passing thunderstorms, but oc-

Temperature Extremes and Records for the Southwestern States

STATE	RECORD HIGH	DATE	LOCATION	ELEVATION (feet)
Arizona	128°F	June 29, 1994	Lake Havasu City	505
California	134°F	July 10, 1913	Greenland Ranch (Death Valley)	−178
Nevada	125°F	June 29, 1994*	Laughlin	680
New Mexico	122°F	June 27, 1994	Waste Isolation Pilot Plant (26 miles southeast of Carlsbad)	3,418
Utah	117°F	July 5, 1985	Saint George	2,761

Source: National Climatic Data Center

* at other locations at the same time or on earlier dates

Saint George, Utah, is the fourth highest location to hold a state's all-time maximum temperature. Only the Colorado and Wyoming records were established at higher elevations: 118°F at Bennett, Colorado (5,484 feet), and 114°F at Basin, Wyoming (3,840 feet).

Landscape north of Yuma, Arizona, near the Kofa National Wildlife Refuge on September 5, 1998
—Sean Potter photo

casionally the desert town receives a year's worth of precipitation from a single cloudburst. These events typically occur in association with a dissipating Pacific hurricane or tropical storm moving north out of Baja California. While precipitation is always welcome, a deluge of 3, 4, or even 5 inches in a day can be disastrous.

On August 9, 1989, a powerful thunderstorm delivered one of Yuma's greatest twenty-four-hour rainfall measurements. The 5.82 inches of rain that fell measured almost twice the city's normal annual precipitation, and resulting flash floods caused severe damage. A similar event occurred on August 16, 1977, when 6.45 inches of rain fell on the city as the remnants of hurricane Doreen moved across the state; coincidentally, this occurred during the El Niño of 1977–78. And on September 17, 1963, Yuma received its most intense rainfall since 1909 as Tropical Storm Katherina dumped 2.43 inches of rain in three hours. In 1989, a thunderstorm on July 27 dropped 2.55 inches, and then just thirteen days later, on August 9, another storm dropped 3.41 inches for a two-week rainfall total of 5.96 inches.

The hottest time of the year in Yuma begins in early June and lasts until late September, when temperatures during the day almost always pass 100°F. Yuma's typical summer daytime temperatures reach 110°F. On September 1, 1950, the temperature in Yuma reached a sizzling 123°F. (Phoenix, another hot southwestern city, measured its all-time hottest temperature on June 26, 1990, when the temperature at Sky Harbor Airport climbed to 122°F. The heat forced officials to close the airport for several hours due to severe turbulence over the airport. Yuma ranks as the fourth hottest U.S. city, with an average annual temperature of 73.9°F; however, based on the average annual *maximum* temperature—87.9°F—Yuma ranks as the hottest U.S. city. With its astounding 4,133 hours of sunshine, meager 3 inches of rain, and numerous days over 100°F, Yuma is the largest U.S. city to endure such extreme desert conditions.

Wintertime Highs

Although the desert Southwest commonly claims the nation's extreme daytime high temperature in the summer, the claim to wintertime extreme high temperatures usually belongs to Florida or southern Texas.

Daily Temperature Extremes: Arizona and California

In a country as large and climatically diverse as the United States, one wouldn't expect the nation's hottest and coldest temperatures to occur in the same state on the same day. That's what typically happens, though, a few days each year when both the hottest and coldest temperatures in the contiguous United States are recorded in Arizona or California—the only two states honored with such temperature extremes. The abrupt changes in terrain—characterized by high mountains and plateaus, steep canyons, and low-elevation basins—and diverse microclimates predispose these states to enormous temperature swings. In addition, the very dry air in these states makes them susceptible to rapid heating during the day and rapid cooling at night. It takes more energy to heat and cool moist air, so humid locales experience less temperature variation than dry locales.

Meteorologists typically record the country's hottest summer temperatures in Arizona (Lake Havasu City or Bullhead City) or California (usually Death Valley), the hottest winter temperatures in Florida, and the hottest spring and fall temperatures in a mix of these states. The country's coldest summertime daily temperatures typically occur in the cold valleys of Wyoming, California, Colorado, Montana, Idaho, and sometimes Oregon. Specifically, West Yellowstone, Montana, Bodie, California, and Fraser, Colorado, claim the dubious distinction of experiencing the nation's lowest daily summertime temperature most often, with roughly twenty to thirty such measurements each year. In 2000, however, Bodie—located at a high elevation (6,671 feet)—reported the nation's (lower forty-eight) low on 113 days. That was far more than West Yellowstone, which most frequently reports the nation's low. In 2000, West Yellowstone only reported twenty-six record lows. During winter, the "iceboxes" in North Dakota and Minnesota usually hold the coldest daily wintertime temperatures.

About two to five times each year, weather observers record both the nation's daily high and low temperature extremes in Arizona or California, for the lower forty-eight. For instance, on June 29, 1997, the temperature rose to 109°F at Casa Grande, Arizona, and plummeted to 28°F at Grand Canyon, Arizona—a difference of 81°F across only 396 miles. And on April 26, 1997, Springerville, Arizona, high on the Mogollon Rim, reported the nation's low of 21°F, while Kingman, in western Arizona, recorded the nation's high of 101°F. On July 6, 1997, the temperature in Blythe, California, reached 115°F while to the north, Truckee, California, reported the nation's low of 32°F—a difference of 83°F across 503 miles. Typically, daytime summer temperatures in the low-elevation deserts (less than 1,500 feet) warm into the mid-100s, while low temperatures in the mountains (elevations greater than 6,000 feet) drop to near 50°F; this results in a normal temperature range of about 55°F across short geographic distances.

Who Determines What's High and What's Low?

The National Weather Service's Hydrometeorological Prediction Center in Camp Springs, Maryland, compiles the national high and low temperatures for the contiguous forty-eight states. Experts determine the daily high and low temperatures using information provided by local National Weather Service offices. A temperature must fit certain criteria for consideration as a national extreme. The temperature must have been recorded in the contiguous forty-eight states at an elevation under 8,500 feet, and the location must have a population of more than 1,000 people. These guidelines represent temperature extremes the public actually experiences. Exceptions occasionally are made for particularly noteworthy sites, such as Death Valley, when the temperature is over 120°F.

The Grand Canyon with a dusting of snow on its rim —Jon Sullivan photo (www.pdphoto.org)

Such temperature differences can occur on most summer days in the Southwest; however, the temperature difference between the lower elevations in the Grand Canyon and the rim above can be remarkable. Take, for example, the extreme case of Inner Canyon, elevation 2,940 feet, and Bright Angel Point (about 1 mile south of North Rim, Arizona), at about 8,148 feet—almost a mile higher and only 5 miles away. In July, Inner Canyon experiences a normal high temperature of 106°F, while Bright Angel enjoys afternoon temperatures in the upper 70s. A similar contrast in temperatures exists in winter as well.

A record of the national high and low in the same state in one day does not happen often. More commonly the same state records the national high and low during the same year. Arizona, California, Colorado, Idaho, Montana, Nebraska, Nevada, New Mexico, Oregon, South Dakota, Utah, and West Virginia have all recorded the nation's maximum and minimum daily temperatures during the course of a single year. In 2000 California recorded both the national high and national low on the same day seventy-five times!

Perhaps the rarest combination occurs when one weather station records both the nation's high and low temperatures during the same year. This has happened only three times to date: Elk City, Idaho, was the nation's hot spot on August 26, 1986, with a high temperature of 106°F and shared the nation's low of 14°F with two other locations (unknown) on October 18; Glasgow, Montana, recorded the highest temperature in the contiguous United States on June 5, 1988, with 108°F, and the lowest on November 17 with −7°F and on December 24 with −6°F; Valentine, Nebraska, has recorded the nation's high and low twice!

Extreme Evaporation: Mojave Desert

The largest and driest deserts of the world lie near two distinct latitudinal belts around the earth. The global circulation of air (see *Introduction*) creates bands of sinking air at 30 degrees north and south latitudes. High pressure, dry air, and limited precipitation characterize these bands, sometimes called the *subtropical ridge.* Most of the world's deserts occur along these bands. This global circulation places the deserts of the Southwest under the influence of stable air masses since the high pressure in the upper atmosphere helps to keep storms away, and sinking air typically warms, further stabilizing the atmosphere. With limited moisture and sparse vegetation to absorb it, the majority of the sun's energy heats the surface beneath this high-pressure system, which in turn heats the air above it. Hot air rising off the surface induces a low-pressure system that heats a large portion of the lower atmosphere. All these factors help create the desert landscape, where rain falls infrequently, humidity stays low, and extreme evaporation occurs.

The Mojave Desert—one of the world's great deserts—spans parts of California, southern Nevada, Arizona, and Utah. It has the most extreme desert weather and climate in the United States and is one of the driest places in the northern hemisphere. The Mojave receives a meager 5 inches or less of precipitation per year. And within the Mojave sits the headquarters of Death Valley National Park, elevation 194 feet below sea level. It receives an average annual precipitation of a mere 1.90 inches. Lying in the local rain shadow of the Panamint Range, Death Valley sometimes experiences no annual precipitation.

The global wind pattern adds insult to injury in the Mojave. Extremely dry because of the rain shadows the Sierra Nevada and Inyo Mountains cast over it, the region's towering mountains wring almost every drop of moisture from once-saturated Pacific storms and leave thirsty deserts virtually no precipitation. The only significant dampening the Mojave Desert receives comes from thunderstorms that the summer monsoon season produces when moisture from the Gulf of California and Gulf of Mexico blows northward.

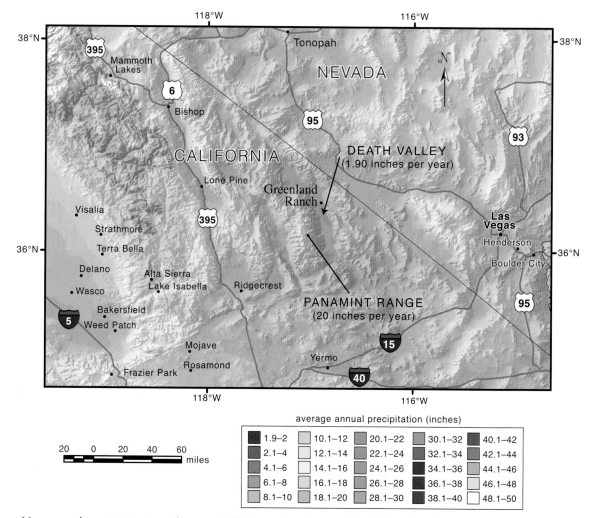

Mean annual precipitation in southeastern California and southwestern Nevada based on the interpolation of average total annual precipitation measured between 1961 and 1990 at weather stations around the states. The interpolation of the weather-station data was done with the PRISM (Parameter-elevation Regressions on Independent Slopes Model) Climate Mapping Program, an ongoing effort to produce detailed, high-quality climate maps. (For more information visit www.ocs .orst.edu/prism/data/) —Data courtesy of Spatial Climate Analysis Service, Oregon State University; illustration by author

Lowest Average Annual Precipitation

The area south of the Arizona-Mexico border is the driest place in North America. According to the National Climatic Data Center, the small Mexican town of Batagues, located at 16 feet above sea level at the northern end of the Gulf of California, has the lowest average annual precipitation in North America. During a fourteen-year period, Batagues averaged only 1.2 inches of precipitation. This compares to the driest spot in the world, Arica, Chile, where over a fifty-nine-year period, the average annual precipitation amounted to only 0.03 inch. During most years it does not rain in Arica, Chile.

Other than that, infrequent winter rains occur as Pacific storms move across northern Mexico, wrapping moisture from the Gulf of Mexico around their counterclockwise flows and sending it northward. Some desert flowers that bloom in the spring rely on winter moisture. The limited summertime moisture from scattered afternoon thundershowers often only refresh small areas. Bagdad, California, in Death Valley, went 767 days—a whopping 2.1 years!—without measurable precipitation; that is, precipitation greater than or equal to .01 inch. Between October 3, 1912, and November 9, 1914, not a drop of rain fell, making that the longest rainless period for any location in United States.

Not only is precipitation scarce in the Mojave but also water vapor is nearly absent from the air. The lowest relative humidities in the United States occur in the Mojave. The relative humidity typically approaches 5 percent. Las Vegas, Nevada, ranks as the United States' least humid city, having an average annual relative humidity of only 30.5 percent. Las Vegas is also the second sunniest city in the United States, enjoying sunshine 85 percent of the year.

The Mojave's abundant sunshine, hot temperatures, low relative humidity, and a number of other factors equate to extreme potential evaporation—"potential" because there isn't a lot of water to evaporate. Evaporation is a measure of the amount of water lost to the atmosphere through the transformation of liquid water to water vapor. Meteorologists sometimes refer to evaporation as *pan evaporation* because it measures how much water is lost from a Class-A pan, which is 4 feet in diameter and 10 inches deep. It is the official device U.S. weather observers use to measure evaporation.

If the Mojave did have moisture to evaporate, the desert's harsh, arid conditions would transform more than thirty times the area's average annual rainfall into water vapor a year. The Mojave measures an amazing 130 to 150 inches of pan evaporation a year—more than anywhere else in the United States. This compares to annual evaporation rates of 55 inches in Salt Lake City, 30 inches in Seattle, and 20 inches in the rain forests of western Washington's Olympic Mountains. On a typical summer day, more than 0.60 inch of water can evaporate from a pan in

Lowest Average Annual Precipitation Extremes in the World

CONTINENT	AVERAGE ANNUAL PRECIPITATION (inches)	LOCATION
South America	0.03	Arica, Chile
Africa	less than 0.1	Wadi Halfa, Sudan
Antarctica	0.8	Amundsen-Scott South Pole Station
North America	1.2	Batagues, Mexico
Asia	1.8	Aden, Yemen
Australia	4.05	Mulka (Troudaninna), South Australia
Europe	6.4	Astrakhan, Russia
Oceania	8.93	Puako, Hawai'i
(the islands of the central and South Pacific, including Melanesia, Micronesia, and Polynesia)		

Source: National Climatic Data Center

the Mojave Desert. The evaporation season peaks in late June, then decreases in July and August as the monsoon sets in. Evaporation data aids in determining water loss from reservoirs, managing crop irrigation, and water balance research for weather forecast models.

Since evaporation from a relatively small and warm surface, such as an evaporation pan, is higher than that from a cool lake, climatologists multiply evaporation measurements from a Class-A pan by 0.65 to give them a more accurate measurement of lake evaporation in a region. Although this correction factor reduces the evaporation off Mojave water bodies from 150 inches to 98 inches a year, it still represents a very significant amount of water loss, especially from reservoirs such as Lake Mead. For Lake Mead, that equates to roughly one-third of its capacity. That's water that cannot recharge the water table or nourish desert flora and fauna, and this, among other factors, makes Lake Mead's existence controversial.

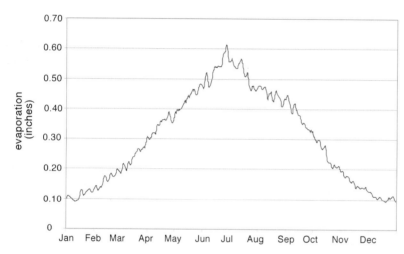

Mean daily evaporation at Boulder City, Nevada
—Courtesy of the Western Regional Climate Center

Aerial photo of Lake Mead
—Photo courtesy of the National Park Service

The population explosion in the desert Southwest introduced numerous man-made reservoirs, waterways, golf courses, and nondesert vegetation, all of which increase the volume of evaporation that occurs. Some theorize that increased evaporation increases the amount of moisture available to stray thunderstorms. One thing is certain, evaporation in deserts transforms huge volumes of water to vapor before it can replenish the groundwater supply.

Southwest Monsoon

The deserts of the southwestern United States experience a monsoon, much like the world-famous monsoon that hits India. *Monsoon* comes from the Arabic word *mausim,* which means "a time" or "a season." Today we use the word when talking about seasonal wind shifts. The Southwest monsoon, a large-scale wind shift, occurs across the desert Southwest and Mexico once a year. Sometimes called the "Mexican monsoon," it is one of the largest seasonal climate signals in North America. People both celebrate and fear the onset of the monsoon, as it can bring welcomed rains but also disastrous floods.

During most of the year, upper-level winds blow from the west or northwest across California and Nevada and into Arizona and New Mexico. These winds lose most of the moisture they once carried across the Pacific Ocean in the Sierra Nevada and Coast Ranges of the West Coast. This leaves southern Arizona and New Mexico relatively dry for nine months (October through June). For example, Silver City, New Mexico, receives half of its total average annual precipitation of 17.48 inches between October and June, and the other half in the remaining three months.

Three systems work together to bring the Southwest monsoon: the Pacific subtropical high-pressure area, a thermal low over the southwestern deserts, and the Azores-Bermuda high-pressure area. For the monsoon to occur, the Pacific subtropical high-pressure area—a major player in all aspects of weather in the western United States—moves from its usual position over Mexico's Pacific Coast north into the Gulf of Alaska. As the Pacific high migrates north, its typical westerly winds loosen their drying grip on the Southwest. At the same time, the Azores-Bermuda high-pressure area over Bermuda moves westward. The Azores-Bermuda high is a large-scale system and its westward movement is key to the monsoon. It is a semipermanent feature in the northern Atlantic Ocean that migrates east and west, and during the summer, it makes it farthest westward movement. This position allows the large-scale clockwise circulation to draw warm, moist air northward into Arizona from the tropics.

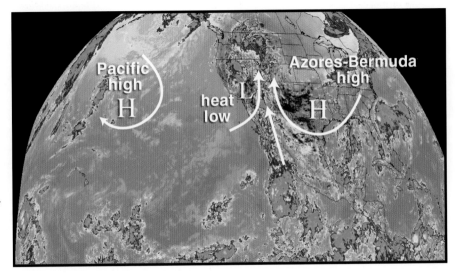

Color-enhanced infrared satellite image on September 2, 1997, showing a strong flow of monsoon moisture streaming northward from the equatorial Pacific. The winds associated with the Azores-Bermuda high over eastern Texas, the heat low over the Southwest, and a weakened Pacific high over the northern Pacific work together to draw moist, tropical air northward into Arizona. The colors denote the temperature of the cloud tops—the yellows and greens indicate the coldest cloud tops and the heaviest precipitation. —Courtesy of the National Oceanic and Atmospheric Administration (annotated by author)

The other key, hot air rising off the Mojave Desert, induces a thermal low (see *chapter 1*) over that region. In combination with the Azores-Bermuda high, the counterclockwise circulation around the thermal low triggers a southerly flow of vast amounts of moist, tropical air from the Gulf of California and the Gulf of Mexico across Arizona and New Mexico. When this happens, residents of the Southwest proclaim, "The monsoon has arrived."

Over Thirty Feet of Rain!

The Indian Ocean monsoon produces one of the most dramatic annual climatic events on the planet. A variety of physical mechanisms produce strong seasonal winds that trigger the characteristically wet summer (and dry winter). As the large Asian continent warms, a heat low, called the *monsoon low*, forms over India. As the low intensifies, it draws warm, moist air from over the warm Indian Ocean. From June to September, the prevailing southwesterly winds bring torrential rainfall. The Indian Ocean Monsoon is particularly dramatic given the vast area of heated land and the warm temperatures of the Indian Ocean. Each summer deadly floods ravage India and Southeast Asia. Cherrapunji, India, holds the world-record rainfall intensity for thirty days and longer. During July 1861, Cherrapunji received 366.14 inches of rain! That amount of rain causes extreme flooding.

Thunderstorms erupt easily and suddenly when the radical increase in available moisture mixes with instability caused by intense surface heating. Officially, the onset of the monsoon at Phoenix, Arizona, is a period of at least three days or more during which dew points average 55°F or higher (typical dew points average 18 to 28°F during the rest of the year). The dew point is the temperature to which the air must cool for the relative humidity to be 100 percent—that is, saturated with moisture—and trigger condensation.

With a temperature of 100°F in the Southwest, a 55°F dew point equates to a relative humidity of 26 percent, which seems low, but feels muggy for Arizona. Since warmer air can hold more moisture than cool air, the Southwest's low relative humidity readings don't accurately reflect the amount of moisture in the air, or the region's potential for thunderstorms. A relative humidity reading of 26 percent in some regions, say the Pacific Northwest, would be too low to spawn thunderstorms because temperatures in that region tend to be cooler than the Southwest, and since cooler air holds less moisture than warm air, a 26 percent humidity reading would mean there was little moisture in the air. Meteorologists recognize that the dew point is a much better indicator of the actual amount of moisture in the air. They use the term *relative humidity* to convey to residents how humid the day will feel.

In order for precipitation-producing thunderstorms to develop, there must be an unstable atmosphere (warm air mass under a cooler air mass), strong uplift, and low-level moisture. Strong uplift is necessary for thunderstorms, but uplift of a lesser degree is often sufficient to produce clouds and light precipitation. When the dew point reaches 55°F in the Southwest, thunderstorms erupt out of thin air because the region's superheated surface air can push the moisture-laden air to heights where condensation can occur. In other areas of the country, a relative humidity reading of 26 percent would not suffice to create thunderstorms since these regions lack sufficient surface heating to create an updraft powerful enough to push air into cooler regions of the atmosphere. For Phoenix, the optimum air temperature for monsoonal thunderstorms is 105°F.

The "wet" monsoon season typically runs from July 1 through September 30. Statistically, August 8 is the peak of the season—when the most rain falls and the most thunderstorms occur—

A cloudburst over southern Utah in June 1988 —Michael Jewell photo

across the southern reaches of Arizona and New Mexico. Typically the monsoon thunderstorms arrive in mid-July and douse the region with 45 percent (central Arizona and New Mexico) to 75 percent (southeast Arizona) of the region's total annual precipitation over a two-month period. The earliest date the monsoon ever hit Phoenix was June 16, 1952, and the latest was July 25, 1987; the average arrival date is July 7.

A period of heavy monsoonal thunderstorm activity, called a *burst,* happens when weak, upper-level storms move across a region and cool the upper atmosphere, increasing the strength and number of thunderstorms. Bursts can devastate desert areas where scarce vegetation and abundant rocks and sand make the area vulnerable to flash floods. Rushing floodwaters easily gouge out roads, homes, bridges, and crops.

A burst in August 1996 brought one of the most severe thunderstorm outbreaks in Phoenix's history. On August 14, a thunderstorm hit parts of northwest Phoenix with strong downbursts gusting to 115 miles per hour at Deer Valley Airport, a state-record wind gust. The wind damage to the city left some without power for several days. Another noteworthy burst came during August 1955, when Tucson received 7.93 inches of rain, making that August the all-time wettest month for this desert city.

When a ridge of high pressure, known as the *subtropical ridge,* forms across northwestern Mexico, it causes a temporary break in the monsoon. This break can occur at any time during the monsoon, but on average the first of potentially several breaks occurs on August 16. The subtropical ridge builds over Mexico as the

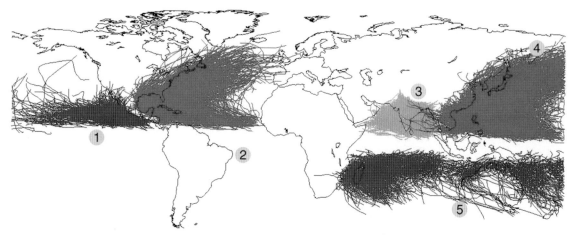

Tropical cyclone tracks for the following time periods: (1) eastern Pacific 1949–2000 (721 storms); (2) western Atlantic and Caribbean 1871–2000 (1,083 storms); (3) Indian Ocean 1877–2000 (1,421 storms); (4) western Pacific 1884–2000 (3,009 storms); (5) south of the equator 1897–2000 (1,293 storms). For plotting purposes the tracks have been truncated at the international date line. —Image courtesy of Richard Knight, National Climatic Data Center

moisture streaming in from the south diminishes, and as natural fluctuations in the subtropical jet stream shift the global highs and lows around slightly. The ridge induces all layers of the atmosphere to warm, which in turn stabilizes it. A stable atmosphere has a difficult time spawning thunderstorms, so it provides a welcome break from the rain.

While the monsoon strongly influences precipitation across Arizona and New Mexico, it also delivers thunderstorm-producing moisture to areas of Utah, Nevada, southeastern California, Colorado, southern Oregon, southern Idaho, and extreme western Texas. Typically, the monsoon moisture surges northward into Oregon, northern Idaho, Wyoming, and Montana a couple of times each summer.

Hurricane Alley of the West

A hurricane, or typhoon, is the earth's largest storm, and it can unleash nearly incomprehensible amounts of energy as wind and rain. The energy a single garden-variety hurricane releases equals that of several atomic bombs. *Hurricane,* derived from

the Tainos word *huracan,* means "evil spirit." The Tainos were native people of the Greater Antilles and the Bahamas. The word *typhoon* is derived from the Cantonese *tai-fung,* which means "great wind." Dynamically, a typhoon and a hurricane are identical, but if a strong enough storm occurs in the western Pacific Ocean, it's called a typhoon, and if it occurs in the Atlantic Ocean, eastern Pacific Ocean, Caribbean Sea, or Gulf of Mexico, it's a hurricane.

Tropical cyclones form hurricanes and tropical storms. For this to happen, several environmental conditions must come together. First, warm ocean water of at least 80°F to a depth of at least 150 feet must supply heat to the cyclone. Second, there must be an unstable atmosphere that cools quickly with altitude, causing moist air to rise. Third, layers near the mid-troposphere (3 miles up) must have moisture. Dry air cannot support thunderstorms, which make up a tropical cyclone, since the energy of rising moist air would dissipate in dry air. Fourth, all these factors must occur at least 300 miles north or south of the equator where the Coriolis force is sufficient to maintain the low-pressure rotation characteristic of cyclones. A tropical cyclone cannot form spontaneously, as can thermal lows, which are generated by intense daily heating. Cyclones require a weakly organized area of low pressure to form. A weak low is necessary at first so the heat can concentrate in a vertical area and not be mixed throughout the atmosphere. And lastly, for a tropical cyclone to form, there can't be a great deal of vertical wind-shear between the surface and the upper troposphere. Large values of vertical wind-shear—the magnitude of wind change with height—disrupt the circulation of tropical cyclones. In fact, large vertical shear, which is common in winter and spring, can weaken or destroy tropical cyclones. When all of these conditions meet, a tropical cyclone can develop. As the cyclone becomes organized, an "eye" develops, which denotes the center (lowest pressure) of the cyclone.

Depending on its strength, a tropical cyclone can become a tropical storm or a hurricane. This occurs most often during the Pacific hurricane season, which runs from May 15 through November 30 and peaks in early October. At this time of year the sea surface temperatures warm enough to initiate and sustain these storms and wind shear is not as strong since the subtropical jet stream has slowed down.

On average, eighteen tropical storms develop a year in the warm waters off the west coast of Mexico; about half of those strengthen into hurricanes—storms that produce winds greater than 73 miles per hour. The general east-to-west flow of winds in the tropics carries most tropical storms westward into the open Pacific Ocean and toward Hawai'i, but most die before striking the islands. A few make it to Hawai'i before weakening. An even smaller number of tropical storms track northward from the west coast of Mexico through Baja California or along the Pacific Coast to blast southern California and southern Arizona—the so-called Hurricane Alley of the West.

Name That Hurricane

Meteorologists classify tropical storms that produce sustained winds of 74 miles per hour or greater as typhoons if they occur over the northwest Pacific Ocean west of the international date line, and as hurricanes if they occur over the north Atlantic Ocean, the northeast Pacific Ocean east of the date line, or the south Pacific Ocean east of 160 degrees east latitude. The National Hurricane Center in Miami monitors and forecasts tropical storms for the eastern Pacific, the area from the Pacific Coast west to 140 degrees west longitude. The Central Pacific Hurricane Center in Honolulu handles storms from 140 degrees west to the international date line.

The names for storms in the eastern Pacific, like the names of Atlantic storms, come from six lists that rotate annually. No names beginning with Q, U, X, Y, or Z are used because such names are rare. In some seasons, the eastern Pacific experiences more than twenty-one named storms. In those seasons, when meteorologists have more hurricanes to name than letters in the alphabet, they add a letter from the Greek alphabet (alpha, beta, etc.) to the hurricane names that start with a W. To name central Pacific storms, meteorologists refer to a list of Hawaiian names that remains in use every year. To avoid confusion, the names of the most extreme hurricanes—well-known storms that caused extensive damage and death—are retired.

Tropical storms and cyclones need warm water with high evaporation rates to survive: water must reach about 80°F to spawn and sustain such events. The cold current off the southern California coast typically hovers in the 50s and low 60s, cold enough to quickly kill any hurricane or tropical cyclone or storm that ventures near. The Gulf of California, however, provides warm water to feed tropical cyclones (and tropical storms when they make it that far north) as they spin northward. Unlike the cold waters that well up off the southern California coast, the Gulf of California's waters can approach 90°F in summer. Surprisingly, the remnants of hurricanes and tropical storms lash southwestern Arizona more frequently than they strike San Diego or Los Angeles.

While a true hurricane-strength storm has never hit the Southwest, at least one tropical storm and the remnants of several hurricanes have. The desert town of Yuma, Arizona, only 70 miles from the Gulf of California, has endured two tropical storms and more remnants of tropical storms and hurricanes than any other town in the West. And although tropical systems typically die quickly after moving over the Pacific's cold water or ashore, even their remnants can hammer the Southwest with severe rain and wind. A large percentage of Yuma's annual precipitation of 3.01 inches comes from tropical storms, and a single tropical storm can deliver all of Yuma's precipitation for the year.

On September 25, 1997, El Niño–aided Hurricane Nora made landfall on the Pacific Coast of Baja California near Punta Eugenia as an 85-mile-per-hour hurricane (down from 130 miles per hour). The next day, Nora blew into Yuma as a weak tropical storm—a rare case of a Pacific system maintaining tropical-storm strength upon entering the United States. Sustained winds of tropical-storm force, 37 to 73 miles per hour, were recorded at Yuma. Nora drenched Yuma with 2.30 inches of rain and dumped almost 1 foot of rain in the Harquahala Mountains in southwestern Arizona.

The drenching rains associated with tropical storms can trigger catastrophic flooding in the desert Southwest, where little vegetation holds the sandy soil in place. Yuma's worst whipping came from Tropical Storm Kathleen—one of only two tropical storms to retain power as far north as the Southwest. Meteorologists had just downgraded the storm from a hurricane to a tropical storm when it raced over the Mexican border near Calexico, California, and swept across Death Valley and into western Nevada on September 10, 1976. Heavy rains and winds up to 76 miles per hour tore into Yuma and caused $2 million ($6.3 million in 2002 dollars) in damage. In places, up to 10 inches of rain flooded the deserts of southern California, which normally receive only 3 to 6 inches of rain a year. The heavy rain sent a 15-foot wall of water into Ocotillo, California, engulfing the town and killing six. When all

Saffir-Simpson Hurricane Intensity Scale

CATEGORY	MAXIMUM SUSTAINED WIND SPEED (mph)	MINIMUM SURFACE PRESSURE (inches)	STORM SURGE (feet)	DAMAGE
tropical depression	less than 37	—	0–2	minimal
tropical storm	37–73	—	1–3	minimal
1	74–95	28.94 or more	3–5	slight
2	96–110	28.50–28.93	6–8	moderate
3	111–130	27.91–28.49	9–12	severe
4	131–155	27.17–27.90	13–18	major
5	more than 155	less than 27.17	more than 18	catastrophic

Source: National Weather Service, National Hurricane Center

Infrared satellite image taken on September 12, 1997, of Hurricane Linda, the strongest tropical storm ever measured in the eastern Pacific, as it spins off the coast of Mexico near 18 degrees north latitude and 110 degrees west longitude, or about 1,000 miles south of Yuma, Arizona. Generating winds estimated at 180 miles per hour, the storm threatened to hit California as a tropical storm but turned away, delivering only high surf and added moisture for showers and thunderstorms. Colors indicate the temperature of the cloud tops; orange and yellows denote the coldest cloud tops and heavier rainfall. Symmetrical bands of thunderstorms around the eye indicate a well-organized and powerful storm. —Image courtesy of the National Oceanic and Atmospheric Administration/National Weather Service

High-resolution image of Tropical Storm Claudette (formerly a hurricane) on July 16, 2003, spinning over the west Texas–Mexico border. This storm dropped 5 to 10 inches of rain. —Image courtesy of the MODIS Rapid Response Team at NASA GSFC

was said and done, Tropical Storm Kathleen had dumped heavy rain on every western state. The National Weather Service named the storm the tenth top weather, water, and climate event of the twentieth century in Arizona.

Another noteworthy storm battered the Southwest between September 4 and 6, 1970. Tropical Storm Norma, centered over the Pacific Ocean south of Baja California, threw tremendous winds, rain, and clouds across Arizona. The surge of moisture exploded into Arizona's deadliest flooding in history. The death toll in Arizona alone reached twenty-three, and twenty-five people in all perished across the Southwest. The worst flooding ravaged central and south-central Arizona, washing out roads, bridges, homes, and millions of dollars worth of crops. A rare tornado—spawned by a Norma thunderstorm—even touched down briefly in Scottsdale on September 5. Even though most hurricanes and tropical storms never make landfall in the Southwest, they can still cause tremendous damage.

The remnants of tropical storms visit southern Arizona about once every five years, nearly as often as similar storms strike some cities along the Florida Gulf Coast. However, sections of southern California, southwestern Arizona, and northern Mexico are not the only inland areas in the West directly affected by tropical storms and hurricanes. On occasion, the remnants of Pacific hurricanes also threaten Colorado, Utah, New Mexico, and Nevada with flooding rains. A couple of these remnants hit New Mexico and Colorado in August 1886 and in August 1970, causing tremendous damage.

No hurricane has made landfall on the California Coast since records began, but September 1939 was very unusual. Four tropical cyclones walloped southern California that month, including the only known tropical storm to move onshore in southern California or the entire U.S. Pacific Coast during the twentieth century. On September 25 that year, a deadly cyclone traveling to the northeast made landfall at Long Beach at tropical-storm strength, generating 50-mile-per-hour winds. The storm dumped more than 5 inches of rain in the Los Angeles basin, 6 to 12 inches in the surrounding mountains, and heavy rain in the deserts. In a six-hour period the storm dropped 6.45 inches of rain on the desert town of Indio, California. Forty-five people died because of flooding, and many more perished at sea.

Only once has the remnant of a Pacific typhoon reached the western United States. The remnants of typhoon Frieda struck the Pacific Northwest with winds of up to 170 miles per hour on October 12, 1962. The storm, which became known as the Columbus Day Storm, killed twenty-three people and caused millions of dollars in damage (see *chapter 1*).

Only rarely do the remnants of Atlantic hurricanes reach the western United States, and when they do, they normally affect only parts of New Mexico, Colorado, and adjacent states to the east. Only two intact Atlantic tropical depressions—formerly hurricanes—have ever made it into New Mexico. The first hit on August 21, 1886, when an unnamed hurricane traveled from the Gulf Coast to east-central New Mexico—weakening to a tropical storm, then to a tropical depression—in ten days. Hurricane Celia, a six-day-old Atlantic hurricane, maintained tropical depression strength as it moved over El Paso, Texas, and into New Mexico on August 5, 1970.

Horrible Haboobs: Arizona

The word *haboob* is derived from the Arabic word *habb* (a word with many spelling variations), which means "violent wind." Haboobs, also dubbed "black blizzards" or "dusters," are devastating dust storms that ravage the world's deserts. They hit mainly in northern Sudan, which suffers about twenty-four haboobs each year, but they also strike the southwestern deserts of the United States, particularly Arizona. In Sudan, haboobs become blinding wind events because of ample sand and dust particles in dry lakebeds. The desert landscape of Arizona provides haboobs ample material to kick up as well.

A haboob is a relatively cold, dense dust storm created by downdrafts that flow out of a thunderstorm. For the classic Arizona haboob to develop, winds associated with a thunderstorm's

A haboob, originating over the Gila River Indian Reservation, rolls north toward the South Mountains in Phoenix, Arizona. —Clayon Esterson photo

California Haboobs

In December 1977, a notable haboob struck California. Winds reached an estimated 192 miles per hour in Arvin, California, and lifted more than 25 million tons of soil from local grazing lands. The wind was strong enough to blow out windows in vehicles and pile sand drifts high enough to plug highways and bury cars. Dust raised by the event dimmed the sun as far north as Reno, Nevada. The haboob killed three people and caused $40 million ($118.7 million in 2002 dollars) in immediate economic losses, which did not include subsequent agricultural losses. The National Weather Service ranks the

windstorm as the ninth top weather event in California during the twentieth century.

On November 29, 1991, a haboob triggered a record 104-car pileup on Interstate 5 near the Coalinga exit in the San Joaquin Valley of California. Winds gusting up to 50 miles per hour whipped up loose sand and dirt from nearby fields, which were experiencing their fourth year of drought, into a haboob that put visibility near zero. The horrible wreck claimed at least seventeen lives and injured more than one hundred fifty people. The California Highway Patrol described the accident as the worst highway pileup in the state's history.

gust front must exceed 62 miles per hour, but weaker haboobs can result from 40- to 50-mile-per-hour winds. Once dust and debris load the air, a haboob is born, and it can travel with a thunderstorm for hours. The winds within the leading edge of a haboob commonly reach 60 miles per hour, but the most severe episodes can unleash 100-mile-per-hour winds.

Haboobs are associated with cool temperatures. Although they generally occur in summer, cool air triggers them. When a thunderstorm releases rainfall into dry air, much of that moisture evaporates before it reaches thirsty desert soils. That evaporative process significantly cools the surrounding air. The cold, dense pool of air subsequently rushes to the ground; this is the downdraft. A downdraft can cool the air temperature by 30°F—or even as much as 60°F—in a matter of minutes. On August 9, 1997, a thunderstorm downdraft cooled Lake Havasu City, Arizona, from a blistering 118°F to a refreshing 72°F in just one hour.

When the cold air hits the ground, it spreads out in front of the storm in what's called a *gust front*. A haboob results when the fierce winds of the front entrain dust, sand, and debris and hurl it

Notable Arizona Haboobs

May 12, 1971: Dust storms suddenly reduced visibility to near zero on Interstate 10 near Casa Grande, which resulted in a chain reaction of accidents involving cars and trucks that killed seven people.

September 5, 1975: Strong winds reduced visibility to near zero in blowing dust and resulted in a twenty-two-car accident on Interstate 10 near Toltec. The accident killed two people and injured fourteen others.

March 3, 1989: A massive dust storm lowered visibility to zero along Interstate 10 in Cochise County. Chain reaction accidents involved twenty-five cars and killed two motorists.

April 9, 1995: Winds with a strong gradient transported a dense cloud of dust that reduced visibility to a few feet along Interstate 10 near Bowie. Blowing dust resulted in four separate vehicle accidents involving twenty-four vehicles. Ten people died, and twenty people received injuries. Observers estimated the high winds blew between 40 and 60 mile per hour.

July 30, 1995: A thunderstorm rapidly developed north of Tucson on the west side of the Santa Catalina Mountains. The storm produced a 60-mile-per-hour wind gust and $1/2$-inch hail just northwest of Tucson. As the storm proceeded northwest, it produced a downburst near the Marana exit of Interstate 10. The haboob, estimated at 60 miles per hour, reduced visibility to near zero. Three separate accidents on both sides of Interstate 10 involved twenty-one cars, injured twenty-four people, and caused about $70,000 in property damage.

July 10, 1997: Downburst winds from nearby thunderstorms kicked up a thick cloud of dust as the storm moved across plowed fields of south-central Arizona. This dust cloud then moved across Interstate 10 between Red Rock and Picacho reducing visibility to zero at times. Twelve collisions involving about thirty vehicles resulted, injuring twenty-five people.

July 6, 1999: A widespread dust storm sharply reduced visibility along Interstate 10 about 7 miles northeast of Casa Grande, Arizona. One motorist died when a series of wrecks knotted a 25-mile stretch of the freeway.

thousands of feet into the air. A frightening wall of dust, a haboob can quickly engulf a city and reduce visibility to a few feet. Dust can extend to heights of 10,000 feet or more in the most severe haboobs. To make conditions worse, turbulent eddies within haboobs can generate vicious dust devils and even tornadoes. Abrasive haboobs strip paint from cars and buildings, defoliate trees, and kill people.

Haboob-producing thunderstorms frequently develop over the Sierra Madre Occidental in northern Mexico. These mountains cause the orographic lifting that initiates these storms. Moving northwest, the thunderstorms race at 20 to 40 miles per hour up the Santa Cruz River valley, southeast of Tucson, and toward Phoenix. As the storms approach Phoenix and the dry, vegetation-limited desert, a billowing wall of dust begins to form.

The sudden onset and blinding nature of haboobs cause numerous traffic fatalities each year in southern Arizona, southeastern California, and southern Nevada. The transportation department has placed remote-controlled signs along haboob-prone highways in southern Arizona to warn motorists when haboobs may appear.

On average, two or three haboobs a year strike Phoenix, but as many as twelve have ravaged the city in a single year. Haboob season runs from May through September (the typical thunderstorm season), with the peak in June. A haboob typically lasts less than three hours, but during those three hours the strong winds can inflict serious damage.

One of the most dramatic haboobs to hit Phoenix struck on July 16, 1971, when the temperature in the leading edge of a thunderstorm dropped 50°F, and the relative humidity rose from 33 percent to 74 percent in a matter of minutes, causing the thunderstorm to grow quickly. Nine unsuspecting motorists died and twenty-seven sustained injuries when visibility dropped to near zero.

Arizona's Biggest Snowstorms

Between December 12 and 20, 1967, two huge snowstorms paralyzed northern Arizona and deposited snow on much of the state. During those nine days, 86 inches of snow fell at Flagstaff. At Winslow, where average annual snowfall is 11.2 inches, 39.6 inches of snow fell. On December 14, a state-record 38 inches of snow fell in 24 hours at the Heber Ranger Station. Snowfall totals over the Mogollon Rim included 102.7 inches at Hawley Lake, 99 inches at Greer, 77 inches at Payson, 46 inches at Prescott, 35.2 inches at Sedona, and 31 inches at the south rim of the Grand Canyon. On December 21, the weather station at Hawley Lake reported a snow depth of 91 inches, the greatest snow depth ever at an official weather station in Arizona. The deserts of southern Arizona did not escape, as measurable snow accumulated at several locations: 84 inches on Mount Lemmon, 27.5 inches at Miami, 17.7 inches at Wilcox, 11 inches at Safford, 5 inches at Wickenburg, 3.8 inches at Douglas, 3 inches at Ajo, 2.5 inches at Gila Bend, and 1.6 inches at Tucson. Many homes, barns, and business structures caved in under the weight of the snow. Eight people—most of them on the Navajo Indian Reservation in northeastern Arizona—died of exposure while trapped on impassable roads or while caught outside for other reasons. The National Weather Service ranks the storm as Arizona's top weather event of the twentieth century.

Desert Snowstorms

Significant snowstorms rarely occur at lower elevations of the deserts (generally less than 4,000 feet) of southern Nevada, southern and western Arizona, southeastern Utah, and southeastern California; however, snow flurries blow through the area once or twice during most winters, and snowfall of an inch or more does accumulate once every four to five years in some of these desert areas. The lowest elevations of southeastern California, though, probably never have snow accumulations.

Three basic weather conditions must exist for snow to accumulate at low elevations in the deserts of the Southwest. First, a slow-moving, cold upper-level low—typically cut off from the main flow of the jet stream—must position itself over Arizona or the extreme southern portion of the Great Basin. This provides the cold air necessary for snow. Second, the upper-level low must be positioned in an elongated trough of low pressure across the Great Basin—extending from the Pacific Northwest to northern Baja California. This elongated trough is important. If it didn't exist, the cutoff low would be south of the jet stream and entirely isolated from the cold air to the north. Third, a strengthening surface low-pressure area,

associated with the upper-level cutoff low, must track across southern Nevada and southern Utah just ahead of the cutoff low. The trough of low pressure, the upper-level low positioned in its base, and the surface low all work in unison, creating a counterclockwise circulation that draws moisture from over the Pacific Ocean into the Southwest. This moisture-laden air wraps around the surface low and combines with the cold air being drawn in from the north by the upper-level low, causing snowfall in the deserts. Ordinarily, high pressure dominates the Southwest's weather pattern during winter, which explains why it doesn't receive a lot of low-elevation snow.

Snowstorms in the desert Southwest typically receive moisture from a moist subtropical fetch—the unobstructed distance air travels over the ocean before reaching land—across the eastern Pacific. Although the subtropical jet stream

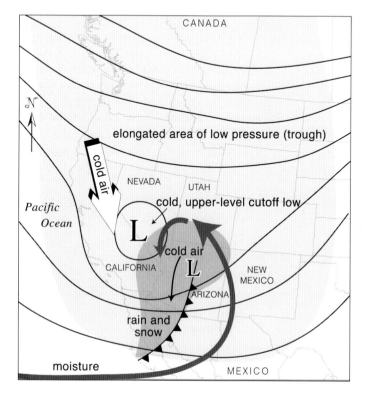

Basic weather elements necessary for snow to occur in the desert Southwest. A cold, upper-level low-pressure system drives southward into the Great Basin and a trough of low pressure develops across the Great Basin. Meanwhile, at the surface, a low-pressure area tracks east just ahead of the upper-level low. The counterclockwise circulation around the lows draws moisture in from the eastern Pacific. As cold air flowing in from the north mixes with moisture wrapping around the low, snow occurs in the deserts.

helps transport the moisture, it plays a minor role in this weather pattern. The amount of energy behind Southwest snowstorms is less important than having the suitable mix of moisture and subfreezing temperatures, which are necessary to produce snow.

When this weather pattern hits the region, cold air blows in from the north or is dragged down to the surface by falling precipitation or the mixing of surface and upper-level winds by a storm. Sometimes the coldest pool of air brought in by the upper-level low is so cold that it moves over the region and destabilizes the atmosphere, promoting embedded thunderstorms. Embedded thunderstorms occur in an otherwise uniform, nonthunderstorm precipitation band. Sometimes these embedded thunderstorms produce thunder snow—a thunderstorm that drops snow instead of rain. Thunder snow usually falls at phenomenal rates—4 inches or more per hour. At the lowest elevations, where the surface temperature is too warm to support snow, heavy showers of precipitation often drag enough cold air from the upper atmosphere to switch the rain to snow at the surface. Although this does doesn't ordinarily result in heavy accumulations, 1 or 2 inches of wet snow can accumulate before the storm moves on and temperatures rise above freezing.

Notable Snowstorms in the Desert Southwest

December 20–21, 1909: *The Las Vegas Age* newspaper headlined a story recounting 10 to 15 inches of snow across the Las Vegas valley from one storm.

March 29–31, 1915: A localized, but heavy, three-day snowstorm buried southeastern Utah. Forty-six inches of snow fell on Moab, Utah (4,000 feet)—the most ever recorded for any month at Moab. After some settling and melting, the snow measured 22 inches deep on March 31.

December 1948–January 1949: A series of storms from December 1948 into February 1949 paralyzed much of Nevada. Las Vegas reported an all-time record monthly snowfall of 16.7 inches in January. The first of three major snowstorms that month came between January 10 and 12, dropping 9.7 inches. That storm still stands as the biggest official snowstorm in Las Vegas's recorded history. Then, on January 19 and 20, a rare two-day snowstorm dropped 2.3 inches of snow. Five days later another 4.7 inches of snow fell. Snowdrifts 12 to 15 feet deep temporarily closed highways in northern and eastern Nevada and Las Vegas received the most snow ever recorded in that city. The National Weather Service ranks the winter of 1948–49 as the number one weather event of the twentieth century in Nevada.

November 16, 1958: Tucson, Arizona, received 6.4 inches of snow.

January 4–5, 1974: Nine inches of snow buried Las Vegas—the second heaviest snowfall in Las Vegas between 1949 and 1999. The monthly snowfall for January 1974 totaled 13.4 inches, making that the second snowiest month on record for Las Vegas.

January 30–February 2, 1979: A total of 7.8 inches of snow fell on Las Vegas, 7.4 inches on January 31 alone.

December 11, 1985: Phoenix received its first measurable snowfall since 1939—0.1 inches.

December 25, 1987: For the first time in forty-seven years, Tucson had a white Christmas as 4 inches of snow blanketed the city.

December 25, 1988: A massive winter storm gave Las Vegas its first white Christmas on record.

Lightning over Tucson, Arizona —© Weatherstock/Warren Faidley

Lightning Capital of the West and Lightning-Ignited Forest Fires

Residents of Tucson like to say their city is the "lightning capital of the world," but that is not entirely true. As a lightning producer, the sky over Arizona ranks far behind Florida, Georgia, Alabama, Kansas, Oklahoma, Colorado, and Wyoming. As is typical around the Southwest, Tucson sees lightning on average forty-two days a year, whereas central Florida experiences more than one hundred days of lightning annually. However, Tucson makes its claim as a mecca for lightning watchers. Each year, the brilliant thunderstorms draw storm chasers and photographers to Tucson from around the world. Perhaps a better title for Tucson would be the "lightning-photographing capital of the world."

The desert's dry air makes the lightning channel in a cloud more visible, thus more vibrant and photogenic. Similar to a light fog, water vapor—suspended water droplets in the air—reduces visibility in humid locations. The lack of clouds and moisture in the desert provides lightning watchers good visibility, even with the presence of dust. In fact, dust enhances the Southwest's beautiful sunsets, which provide backdrops to lightning storms. Beside that, the clouds that do exist in the Southwest are generally so high that they leave plenty of room for lightning to put on a show between the clouds and the ground. The ground elevation in Tucson is about 2,000 feet, while the cloud bases are usually at 10,000 feet or higher, so lightning bolts zigzag through 2 to 4 miles of clear air. Photographs capture the white calligraphy of lightning against the spectacular landscape of the Southwest. Although beautiful, the photographs don't reveal that lightning leads all weather-related events in deaths in the United States.

The National Weather Service reports that direct or indirect lightning strikes kill, on average, 73 people each year in the United States. Other federal sources give higher estimates; the Federal Emergency Management Agency, for instance, reports 150 to 200 deaths and another 750 severe injuries annually due to lightning strikes. Such statistics give lightning the dubious distinction of being the most dangerous force in nature.

Wherever you live in the United States, you are twice as likely to be killed by a lightning strike as by a tornado, hurricane, or flood. The odds of getting hit by lightning sometime in your life are on the order of 1 in 10,000. The odds of a lightning bolt hitting any particular square mile of ground is between 0.05 and 0.80 times the number of thunderstorm days a region experiences each year. So any particular neighborhood in Tucson, which sees about forty-two thunderstorm days a year, could expect as few as two cloud-to-ground strikes or as many as thirty-three a year.

Freakish Lightning Bolts

On September 1, 1939, a single lightning strike killed 835 sheep. The animals had just bedded down for the night high up in Pine Canyon in the Raft River Mountains of Box Elder County in northwestern Utah. Rain from a passing thunderstorm wet the ground and the sheep and allowed the lightning's electrical charge to flow easily through the herd. A similar event occurred on July 18, 1918, when two large bolts of lightning struck and killed 654 sheep on Mill Canyon Peak in the American Fork Canyon in north-central Utah.

Much of the United States' scientific research related to lightning takes place at the University of Arizona in Tucson. Florida, the thunderstorm capital of the United States, also supports extensive lightning research, particularly at the University of Florida's International Center for Lightning Research and Testing at Camp Blanding, Florida. Scientists study lightning for a variety of reasons: to understand its connection

Lightning Types

Lightning comes in two main types: cloud-to-ground lightning and cloud-to-cloud or in-cloud lightning. Cloud-to-ground lightning, the common, negatively charged lightning bolt, travels between the cloud and the ground. While the most common, this type only represents about 20 percent of all electrical discharges

Cloud-to-cloud lightning, also called *intracloud* and *intercloud* lightning, is the most common type of lightning. It is a discharge that travels from one cloud to another, or within the same cloud. Cloud-to-cloud lightning travels between a positively charged center and a negatively charged center, either in the same cloud or in two separate clouds. These flashes usually precede a cloud-to-ground strike. Lightning has many different forms and appearances.

Bolt-from-the-blue or cloud-to-air: Positive lightning, also called the "positive giant," occurs when in clear skies a renegade lightning bolt strikes "out of the blue." Bolt-from-the-blue lightning comes from distant thunderstorms as far away as 10 miles and perhaps not even in sight. A bolt travels horizontally away from the top of the parent cloud, then turns down toward the ground and strikes violently.

Heat lightning: Flashes that make the sky appear orange. Like sheet lightning, these flashes are created by lightning bolts in thunderstorms more than 10 miles away. Since these flashes occur in the hot summer, they are called heat lighting. In the summer, the air is often filled with dust particles that refract the light of distant lightning and cause the bolts or flashes to appear orange. Flashes of heat lightning are too far away for an observer to hear thunder.

Sheet lightning: Clouds that flash and look like luminous white sheets. The clouds appear this way because lightning flashes inside them or behind them.

Ribbon lightning: Ordinary cloud-to-ground lightning that appears to spread horizontally in a ribbon of parallel luminous streaks even though it does strike the ground. Occurs when a very strong wind blows at right angles to an observer's line of sight.

Ball lightning (Saint Elmo's Fire): Also called "globe lightning," these mysterious and rare luminous spheres appear to float in the air or slowly dart about for several seconds. Ball lightning often appears in the vicinity of thunderstorms or a recent lightning strike, suggesting it's electrical in composition or origin. Due to lack of physical evidence, ball lightning continues to baffle scientists.

Bead lightning: When the lightning channel of a cloud-to-ground strike breaks up, or appears to break up, the lightning bolt leaves behind beads of light that appear to be strung together by lightning. Scientists don't know what causes bead lightning.

Forked lightning: Crooked or branched channels in a lightning stroke.

High-altitude lightning: Recently discovered, these colored flashes of light high above thunderstorms have received playful names such as "red sprites," "blue jets," and "green elves." They shoot up from tops of thunderstorms at about the same moment lightning discharges within a storm cloud. Scientists have just begun to collect data on the cause of high-altitude lightning, and have no definitive answers.

to severe weather and the dangers involved with lightning strikes, to evaluate criteria for launching space shuttles, and to learn how to detect and predict lightning better.

Scientists at the University of Arizona in Tucson use a high-tech research facility on a nearby National Guard base to study lightning. At the base in the open desert safely outside Tucson, researchers fire modified rockets into thunderclouds to trigger lightning. To induce lightning, researchers have created a high-tech version of Ben Franklin's kite experiment. They attach a spool of copper wire to a small rocket (about 4 feet high) and affix one end of the wire to the launch pad. When a thunderstorm passes overhead and lightning threatens, the scientists fire the rocket into the sky, about 2,000 feet up. As expected, the lightning takes advantage of the direct connection to the ground. The current travels through the wire to the researchers' monitoring equipment. From these strikes, researchers have gathered a great deal of knowledge about the dynamics of lightning.

Lightning develops when opposite electrical charges build within the cumulonimbus cloud of a thunderstorm. Scientists theorize that negative flash lightning—a type of cloud-to-ground strike—occurs when the ice crystals in the upper part of the cloud carry a positive charge, while the heavier water droplets in the bottom of the cloud carry a negative charge. As the storm cloud moves over the earth, it causes a positive charge to form near the ground. The opposing electrical charges—negative at the cloud's base and positive at the earth's surface—grow in strength and attraction to each other. Eventually, the insulating layer of air between the two cannot keep them apart, and a discharge of electricity in the form of lightning takes place.

First, a rush of negatively charged electrons rushes to the base of the cloud and then toward the ground in a series of steps producing a zigzag pattern. This is called the *stepped leader;* it is very faint and usually invisible to the human eye. As the stepped leader approaches the ground, the potential gradient—the difference between the negative charge traveling down from the cloud and the positive charge of the earth—increases. A current of positively charged protons from the ground, usually from elevated objects, travels up toward the stepped leader; eventually they meet. After they meet, large numbers of electrons flow to the ground and a much larger and more luminous positively charged *return stroke* (lightning) surges up to the cloud, following the stepped leader's path. The return stroke travels extremely quickly—about 60,000 miles per second—and takes only a fraction of a second to reach the cloud. Each leader commonly has several return strokes associated with it, which makes the lightning flicker and flash in the same place several times.

Thousands of Flashes per Hour

A mesoscale convective system, or MCS to meteorologists—a massive group of thunderstorms—can produce lightning at the astonishing rate of several thousand flashes per hour.

While a brilliant bolt of lightning may appear to be a huge, thick column, it has only the diameter of a pencil. Lightning appears thick because the light emitted from the electrical discharge shines brightly. The lightning bolt superheats the air surrounding it, which causes the air to expand and then contract rapidly. The air around the bolt exceeds 40,000°F, which is hotter than the sun. This rapid change in air temperature and pressure creates sound waves that move outward in all directions—the cracks, booms, and rumbles of thunder. The sound waves then propagate a shock wave—a major disturbance in the air that travels faster than the speed of sound—causing a clap of thunder ten times louder than a chain saw. A person standing within a square mile of a spot where lightning strikes will hear the thunder less than three seconds after the flash, and at that distance, the sound creates a deafening crack. Thunder's pitch, loudness, and form, such as cracking and rumbling, all depend on the nature of the lightning flash and the listener's distance from the bolt. The loudness and duration of thunder depends on the energy of the lightning's electrical current: the more energy, the louder and longer the thunder. The sound of thunder rarely carries more than 10 or 15 miles.

Central Florida's "lightning alley" has the highest frequency of lightning in the United States, and with so many opportunities for year-round outdoor activities, it is the most likely state in which to get hit by lightning. According to the National Weather Service, the United States sees an average of 73 lightning-related deaths each year, but some researchers believe this statistic is underreported. In Florida alone, an average of ten people die each year due to lightning; however, Colorado leads the West in lightning-related deaths with an average of two per year. In Arizona, Pima County leads the state in lightning fatalities and injuries, mainly due to the fact that most of the county's population lives close to mountains. The mountains in Pima County trigger a lot of thunderstorms.

Fire Weather Forecast Terminology

A Red Flag Warning or Alert is issued when the National Weather Service predicts severe fire weather, such as low humidity accompanied by strong shifting winds and numerous dry lightning storms. These warnings alert land management agencies and the public of potentially hazardous weather conditions.

A warning goes out when the afternoon and evening relative humidity levels are expected to fall to 25 percent or less; the winds 20 feet above the surface are expected to exceed 15 miles per hour for at least two hours; and ten-hour fuels (fuels that range from 1/4 to 1 inch in diameter and whose moisture levels fluctuate more readily with atmospheric conditions) have 8 percent moisture or less and are expected to remain at that level for two or more days.

During a Red Flag Warning, the U.S. Forest Service can implement and enforce a strict fire ban in national forests. A Red Flag Watch alerts the public of a possible Red Flag event in the near future.

The Coal Steam Fire burns out of control in Glenwood Springs, Colorado, on June 8, 2002. The monument commemorates fourteen firefighters lost in the 1994 Storm King Mountain Fire. —Photo courtesy of Bryan Dahlberg, Federal Emergency Management Agency

Cloud-to-ground lightning sparks the vast majority of the Southwest's forest fires. Lightning ignites approximately ten thousand forest fires each year in the United States and the resulting fires have caused upwards of $300 million in damage annually. The average annual cost of fire suppression between 1999 and 2002 was $1.1 billion. Lightning causes half—and in some years more—of all the forest fires in the United States. Lightning ignites more forest fires in the mountains of the West, including the Southwest, than anywhere else in the United States. In general, the average annual number of lightning-ignited fires in the Southwest is ten to twenty per million acres; however, western New Mexico and eastern Arizona experience sixty to seventy per million

acres (another regional high of twenty to forty per million acres occurs in the Cascades). Notorious for its quantity of lightning-ignited forest fires, the Southwest has what it takes: plenty of fuel and dry thunderstorms.

The average annual precipitation of the Southwest ranges from 4 inches in the low valleys (less than 3,000 feet) of southeastern California to 40 inches in the higher mountains (generally above 7,000 feet) of southwestern New Mexico and eastern Arizona. The precipitation in the mountains supports many types of vegetation, including ponderosa pines, juniper, shrubs, and grasses. However, the desert conditions of this region make this vegetation highly flammable—perfect fuel for lightning-ignited firestorms.

Residents of the Southwest will hear thunder fifty to sixty times a year but only half of these thunderstorms produce rain. Most of the thunderstorms that strike this region are high-based cells that produce spectacular lightning, gusty winds, and cool temperatures. High-based cells have high cloud-bases because of the extremely dry air below them. Any rainfall from these high-based storms evaporates before reaching the ground. Known as *virga*, these evaporating streaks of rain create a cold pocket of air that rushes to the ground and spreads out horizontally as erratic and gusty winds that blow up to 60 miles per hour. Without a dousing of rain, lightning from these storms can easily spark a fire, which can be blown into an inferno by the storm's winds. Often lightning starts a small fire that smolders for a day or two, or even a week, before strong winds fan it into a wildfire.

In some places in the West, the depth of vegetative fuel has built up over the years because of fire suppression. Even though lightning-ignited forest fires naturally cleanse forests of vegetative debris, add nutrients to the soil, and create habitat for some plants and animals, when lightning strikes in these areas with plenty of fuel, calamitous fires can occur.

The Nation's Most Deadly Wildfires

Not all deadly forest fires have occurred in the West. On October 8, 1871, the most deadly wildfire in U.S. history burned Peshtigo, Wisconsin, along the west shore of Green Bay close to Michigan's Upper Peninsula. Embers from burning stumps and rubble exploded into a fire that killed 1,182 people and destroyed over 1.2 million acres of timberland.

The nation's second worst wildfire tragedy happened on September 1, 1894, near Hinckley, Minnesota. A searing heat wave and drought contributed to a fire that killed 418 people and charred 160,000 acres of forest. Small fires associated with unregulated logging practices started the blaze. A southerly wind blowing over these small fires picked up smoking embers and quickly ignited cut-over forests. What locals called the "cyclone of fire" came and went within a few hours.

Significant Lightning-Ignited Wildfires of the West

August 20–September 9, 1910: A massive series of 1,736 forest fires, called the Big Blowup and the Big Burn, in eastern Washington, western Montana, and northern Idaho scorched 3 million acres of timber—mostly in the Bitterroot Mountains of Montana—and killed eighty-five people, seventy-two of them firefighters. The smoke from the fires reached so far that ships in the Pacific Ocean got lost in it. The smoke also reached the Atlantic seaboard, and for the first time, Congress decided to spend federal money fighting forest fires. Abnormally low precipitation, high temperatures, and high-based thunderstorms in the region made August 20 and 21 among the two most grueling days of firefighting history in the United States. The National Weather Service ranks the fires as Washington's fourth top weather-related event of the twentieth century.

August 14–24, 1933: Oregon's second largest forest fire, dubbed the Tillamook Burn, destroyed more than 310,000 acres (13 billion board feet) of dry timber near Tillamook, Oregon. An unusually hot, dry summer set the stage for this immense disaster. Eighteen hundred firefighters initially battled the fire, but ten days later, dry and strong easterly winds fanned the flames to cover 250,000 acres in just twenty hours, making it impossible to battle. Phenomenally powerful, the fire burned huge stands of old-growth Douglas-fir, including some five-century-old trees. The fire was 18 miles long, and it illuminated the night skies for miles. A smoke plume 8 miles high carried ash out to sea where it fell on ships as far as 500 miles away from the coastline. Ash and cinders 2 feet deep covered a 30-mile stretch of beach along the northern Oregon Coast. The fire burned uncontrolled for a week until moist, westerly winds helped decrease its intensity. Officials estimated the total economic loss in excess of $600 million ($8.3 billion in 2002 dollars). The National Weather Service ranks the event as Oregon's fourth worst weather-related event of the twentieth century.

Fall 1988: Several benign lightning-ignited blazes, first started on June 22, quickly flared into an intense firestorm in Yellowstone National Park. Dry and windy weather—one of the driest and windiest summers the park had seen since its opening in 1872—aided the fires. Around 1.4 million acres of the Yellowstone ecosystem—793,880 acres within the national park—came to resemble a lunar landscape of ashen plains. Attempts to extinguish the fire, which eventually consumed 36 percent of the park, resulted in the largest firefighting effort ever waged in the United States. A total of 9,500 firefighters from around the country participated in what came to a $140-million effort ($212.8 million in 2002 dollars). Firefighters built more than 800 miles of fire line, used 117 aircraft, brought in more than 100 fire engines, and dropped more than 1,000,000 gallons of fire retardant. In spite of this incredible effort, they failed to stop any of the fires from running their course. It wasn't until fall, when rain and snow arrived, that the fires were finally extinguished.

July 6, 1994: One of America's worst wildfire-fighting disasters occurred on Storm King Mountain, just 8 miles northwest of Glenwood Springs, Colorado. High winds associated with a cold front and a dry thunderstorm whipped up a firestorm that killed fourteen smokejumpers.

Summer 2000: Eighty thousand separate fires burned nearly 7 million acres across the West; during the course of a normal summer, 106,000 fires burn about 3.6 million acres. A single high-based thunderstorm moved across western Montana and sparked an estimated two hundred forest fires. The 2000 wildfire season was one of the most costly fire seasons ever.

July 13–September 5, 2002: Oregon's largest wildfire, known as the "Biscuit Fire," burned 499,965 acres in southwestern Oregon and northwestern California near the Oregon towns of Brookings, Gold Beach, and Cave Junction. Ignited by lighting, it burned for 120 days. Remarkably, it destroyed only four homes, nine outbuildings, one fire lookout, and numerous recreational structures. At one point, seven thousand firefighters and support personnel fought the fire. The total firefighting cost was estimated at $153 million.

In the Southwest, fire season runs through the summer and into the fall. It peaks during the early fall when electrical storms are most common and vegetation the driest. The peak in storm activity coincides with the Southwest monsoon (see *Southwest Monsoon* earlier in this chapter), a period of relatively moist weather in the Southwest. Typically the atmospheric moisture from the monsoon forms scattered clouds early in the day, and then in the afternoon, superheated air rising off the surface cools and condenses into towering thunderstorms. By evening, thunderstorms begin displays of vivid lighting and heavy rain (from storms that are not high-based). For those areas not moistened by the rain, lightning easily sparks fires in vegetation dried by four months of hot summer temperatures.

Lightning always seeks the shortest distance from the cloud to the ground, so tall structures are most vulnerable to strikes. If caught outdoors in an electrical storm, avoid objects that project above the landscape, such as trees, isolated shelters, and hilltops. Your best option is to crouch in the open, keeping as far away as possible from isolated trees that might be struck. A car or truck provides an excellent place to seek shelter during a lightning storm.

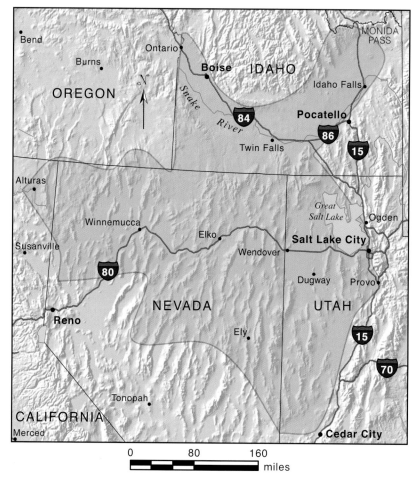

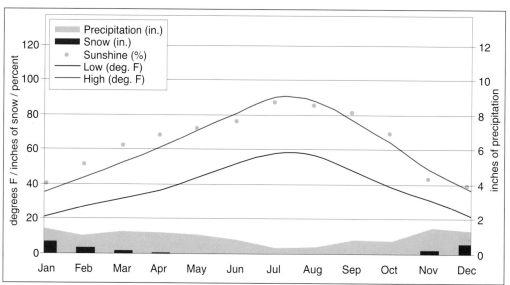

Climate of Boise, Idaho

7

Snake River Plain and Great Basin

Elevation: 5,000 to 13,000 feet

Natural vegetation: grass and other herbaceous plants, scattered needleleaf evergreen trees, sagebrush, and other dwarf shrubs

Köppen climate classification: steppe, semiarid, and arid desert

Most dominant climate feature: generally dry and cool, with dramatic diurnal temperature changes

January and February are the coldest and among the cloudiest months of the year in the Great Basin and Snake River Plain. Some of the cloudiness is associated with inversions that trap fog in the low valleys. Normal daytime temperatures range from the 30s in January to the 40s in February. At night, however, temperatures plummet into the teens and even single digits in the mountains. Total precipitation for these two months ranges from 1 to locally 6 inches, which all falls as snow.

The long bouts of fog and bitterly cold days generally end by March, and beautiful spring weather dominates April. Total precipitation for March and April ranges from 1 to 5 inches, with the greatest amounts in the mountains and in the northern sections of the Snake River Plain. Temperatures can still be cold at night with readings in the teens and 20s, but by April daytime highs warm into the 50s and 60s.

By late May most places have experienced their last spring freeze, and warm conditions begin in June. Average daily temperatures range from the 50s in May to the 50s and 60s in June. Precipitation in May and June is generally less than 2 inches across the region. What precipitation does fall comes in the form of showers and thunderstorms.

July and August are the hottest and driest months of the year. Generally less than 2 inches of precipitation falls through these months, if any falls at all. Some monsoonal moisture (see *Southwest Monsoon* in *chapter 6*) from the south causes welcomed showers and thunderstorms in southwestern Utah and southeastern Nevada. Normal daytime temperatures rise quickly into the 80s and 90s, only to cool off at the same rate to comfortable 50s and low 60s overnight.

September brings cooler temperatures and a slight increase in precipitation, which continues into October. September's daytime temperatures are generally in the 70s, but they cool to the 60s in October. Strong diurnal temperature changes lead to morning low temperatures in the 30s and 40s in September and into the 20s and 30s in October. The first fall freeze occurs in September in much of the region. Trace amounts of precipitation accumulate in most areas, but some areas can see up to 2 inches locally.

Winter's cold, snowy, foggy, and windy conditions have arrived by November and December. Average daily temperatures drop into the 30s, with highs in the 40s and 50s and lows in the teens and 20s. Precipitation totals range from 1 to locally 4 inches, most of which falls as snow.

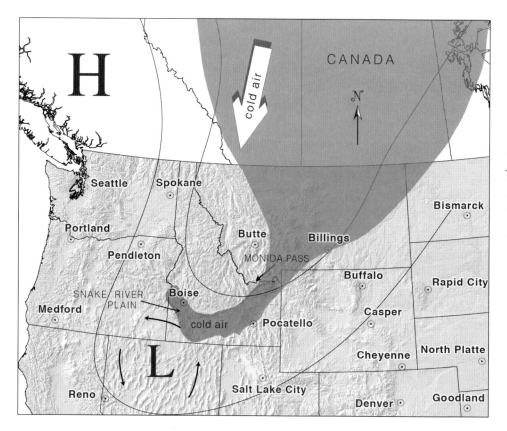

Meteorological setting for a "slosher": arctic air spills into the Snake River Plain through the passes along the Continental Divide, including Monida Pass

Sloshers and Northern Invasions

For most of the winter, the Continental Divide protects the cities of Boise, Pocatello, and Twin Falls from cold arctic air masses to the north. By a process known as *cold air damming,* the Rockies keep the cold air at bay. But every so often the barrier fails, and frigid arctic air pours southward onto the Snake River Plain.

The progress and extent of an arctic outbreak in the Snake River Plain is largely determined by the direction and strength of upper-level winds and surface pressure patterns. Furthermore, a strong correlation exists between snow cover and the low temperatures associated with outbreaks of arctic air—it is virtually impossible to get subzero readings in the Snake River Plain without snow cover.

Seventy-five percent of all arctic air masses that enter the Snake River Plain "slosh" over Monida and other passes situated along the Continental Divide. Monida Pass, where Interstate 15 crosses the Montana-Idaho border, offers one of the largest gaps that arctic air pours through. The arctic air flows through the pass and down the Snake River Plain to Boise, hence the nickname "sloshers." The cold air tends to overtop Monida Pass since the pass is lower than the mountains on either side of it. Meteorologists often can track the leading edge of a slosher as temperatures plummet in one town at a time as the air oozes west and northwest across the Snake River Plain.

On occasion, arctic air will invade the Snake River Plain at its northwestern edge, spilling in through the narrow north-south-oriented valleys of north-central Idaho, southeastern Washington, and northeastern Oregon. These northern invasions account for the remaining 25 percent of arctic air masses that enter the Snake River Plain.

The Boise Mountains and Snake River Plain near Boise, Idaho —Jon Sullivan (www.pdphoto.org)

A great number of outbreaks of arctic air in Idaho have characteristics of both sloshers and northern invasions: cold temperatures, dry air, and sudden onsets. Meteorologists can only tell them apart by mapping the behavior of the jet stream and an upper-level ridge of high pressure that forms over the British Columbia Coast in winter. For sloshers, the ridge must be in south-central British Columbia, thus rotating icy air clockwise from over Canada into western Montana and up against the Continental Divide. The upper-level ridge induces an area of low pressure over the central intermountain region. As a result, a strong pressure gradient develops across eastern Idaho and western Montana. When this pressure gradient is strong enough, it sloshes cold air over Monida Pass and across the Snake River Plain.

On the other hand, a strong ridge of high pressure must form over the southern coast of British Columbia for northern invasions. When this ridge forms along the southern coast, its clockwise flow drives cold air from the frozen prairies of Canada, including the Yukon and the interior valleys of northern British Columbia, into eastern Washington, eastern Oregon, and eventually into Idaho.

Sloshers and northern invasions require snow cover to drive temperatures below zero across the Snake River Plain; without it, the ground can absorb enough heat during the day to counteract the arctic air's nighttime chill. The depth of snow cover does not matter. An inch of snow is nearly as effective as 5 inches of snow; at either depth its insulating qualities reduce the amount of heat the ground can absorb from the sun during the day.

One of the most extreme outbreaks of arctic air to chill the Boise, Idaho, region occurred during the winter of 1985—also one of the area's snowiest years with nearly 20 inches falling in the last two weeks of November. Additional snowfall in early December added to the hefty snowpack, but snow stopped falling as the extremely cold air mass settled over the Snake River Plain; Boise had twenty straight days (December 11–30) with subzero minimum temperatures; the coldest reading of –9°F occurred on December 28. Capped by an inversion, the cold, shallow air mass became severely polluted with wood smoke. Boise residents experienced an extremely poor air-quality day on Christmas Day that year. By New Year's Day 1986, the arctic air mass finally mixed out and was pushed northward by mild southeasterly winds. But the wind died and couldn't keep the cold air out of the valley, so it roared back into the Snake River Plain at speeds of 58 miles per hour.

While severe, the 1985–86 arctic blast is not the coldest on record. The coldest outbreak this century at Boise occurred during December 1990 when temperatures reached –25°F on December 23. For an entire week the average daily maximum temperature was only 2°F and the average minimum was –16°F. At the height of the outbreak, Boise stayed below zero continuously for three days. Other notable cold winters in Idaho include 1937–38 and 1948–49. January 1949 was the coldest month in Idaho since record keeping began in 1893; many stations felt their coldest average monthly temperatures ever. The average monthly temperature in Boise that month was only 10.3°F.

Winds of the Desert and Wasatch Canyons: Utah

Warm canyon winds that bolt out of the canyons of the Wasatch Mountains and into the Salt Lake City, Provo, and Ogden areas are known locally as "easterly canyon winds" or "Wasatch Canyon winds." They can easily blow at or higher than hurricane strength (74 miles per hour) at canyon mouths. Since they blow out of canyons, the scientific community refers to them as *mountain-gap winds*. Canyon winds, highly localized, erratic, and often sudden, have inflicted millions of dollars of damage on this region of Utah.

Often the windiest areas are near Farmington, Centerville, and East Layton, just north of Salt Lake City, where the winds cause significant damage to Cache, Weber, Davis, Salt Lake, and Utah Counties. The force of the wind decreases dramatically as it travels west away from the steep mountains. West Valley City (western Salt Lake City) may only feel a gentle breeze, while the wind howls to 80 or more miles per hour at Centerville. Only occasionally is Salt Lake City

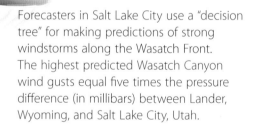

Forecasters in Salt Lake City use a "decision tree" for making predictions of strong windstorms along the Wasatch Front. The highest predicted Wasatch Canyon wind gusts equal five times the pressure difference (in millibars) between Lander, Wyoming, and Salt Lake City, Utah.

SURFACE PRESSURE DIFFERENCE (millibars)	SURFACE PRESSURE DIFFERENCE (inches)	STRONGEST WIND GUSTS (miles per hour)
2	.0591	10
4	.1181	20
6	.1772	30
8	.2362	40
10	.2953	50
12	.3544	60
14	.4134	70
16	.4725	80
18	.5315	90
20	.5906	100
22	.6497	110

International Airport—located 8 miles west of the canyon mouths—buffeted by winds over 50 miles per hour.

Canyon-wind season runs from December through May. In order for canyon winds to occur, there must be a dramatic difference in east-west surface pressure across Utah. A strong pressure gradient forms across Utah when a strong area of low pressure develops over southeastern Nevada and high pressure settles over Wyoming. The pressure gradients mixed with strong upper-level winds force the air over and around the Wasatch Mountains into the low-pressure area in Nevada. The windiest conditions occur with strong

Severe Canyon Windstorm Events in Northern Utah

April 23, 1931: A strong storm system generating fierce easterly foehn winds across northern Utah blew freight cars from their tracks.

September 21–22, 1941: A violent windstorm lashed the Salt Lake City–Ogden area with 50- to 100-mile-per-hour winds. Salt Lake City Airport was rocked by a gust clocked at 80 miles per hour, the strongest wind gust ever recorded at the airport.

May 16, 1952: Possibly the most damaging wind event ever in the Salt Lake City–Ogden area. Eighty- to one-hundred-mile-per-hour winds inflicted an estimated $1 million ($6.8 million in 2002 dollars) in damage to trees and buildings. Winds gusted up to 95 miles per hour at Hill Air Force Base in Ogden.

April 3, 1964: A severe canyon-wind event battered much of the Wasatch Front of northern Utah. Winds gusted up to 57 miles per hour at Salt Lake City Airport, the strongest ever in April. Several trucks blew off Highway 91.

April 4–5, 1983: One of the strongest easterly windstorms in Utah's history blasted the Wasatch Front of northern Utah with winds over 100 miles per hour. Gusts of 103 miles per hour were registered at Ogden and a state low-elevation record gust of 104 miles per hour was recorded at Hill Air Force Base, later eclipsed in April 1999. The fierce winds blew down high-tension power lines, overturned mobile homes,

rolled trucks off Interstate 15, and even overturned twelve flatbed railroad cars near Farmington. Almost every glass window in Ogden shattered. The total damage estimate soared to $8 million ($14.4 million in 2002 dollars). The National Weather Service ranks it as Utah's eighth most significant weather event of the twentieth century.

April 2, 1997: The Wasatch Front was blasted with winds of 50 to 80 miles per hour. The highest winds hit Mount Ogden, 101 miles per hour, and Centerville, 84 miles per hour.

April 23, 1999: On this day the pressure difference between Salt Lake City, Utah, and Lander, Wyoming, reached 19 millibars. At 12:59 AM officials clocked a wind gust of 113 miles per hour at Brigham City Airport. This wind gust established a new wind-speed record for a low-elevation site (below 5,000 feet) in Utah. The previous record had been 104 miles per hour set at Hill Air Force base in 1983.

March 21, 2000: High winds caused widespread damage and toppled nineteen rail cars in Davis County. The maximum recorded gust in the Salt Lake City Valley was 86 miles per hour in Davis County, but the rail car damage indicates winds may have exceeded 100 miles per hour. Peak gusts in the mountains northeast of Salt Lake City included 102 miles per hour at Hidden Peak (also known as Snowbird) and 83 miles per hour at Francis Peak.

northeasterly, easterly, or southeasterly upper-level winds. These upper-level winds flow parallel to the surface winds, enhancing the strength of the surface winds. If the upper-level winds blow perpendicular to the surface winds, they tend to reduce surface wind velocities.

As the strong easterly winds descend the western slopes, compressional heating warms the air, but not to the extent of Chinooks that hit the Rocky Mountain Front. Since the canyon winds develop from air over the frigid Great Divide Basin in southwestern Wyoming, they are very cold to begin with.

A rule of thumb holds that the highest wind gusts in the windiest canyons will equal five times the difference between the surface pressure (in millibars) at Lander, Wyoming, and Salt Lake City, Utah. For example, if the pressure is 1,026 millibars (30.30 inches) in Lander and 1,010 millibars (29.92 inches) in Salt Lake City, then the

pressure difference is 16 millibars (.4725 inches of mercury). This means the strongest wind gusts will reach 80 miles per hour. When the pressure difference reaches 18 millibars (.5315 inches of mercury), the winds will break through the mountain gaps regardless of the direction of the upper-level winds. (The Salt Lake City and Lander weather stations' pressure readings can be easily accessed using a detailed weather map online.)

Barometers measure pressure in inches of mercury. A pressure reading of 29.92 represents how high the pressure would raise the mercury in the barometer, which is a sealed tube. A reading of 29.92 inches of mercury equals 14.7 pounds of pressure per square inch, the average pressure a column of air 10 miles tall exerts per square inch of ground at sea level.

These easterly Wasatch Canyon winds can make for treacherous driving conditions by blowing snow and creating drifts. The shear strength

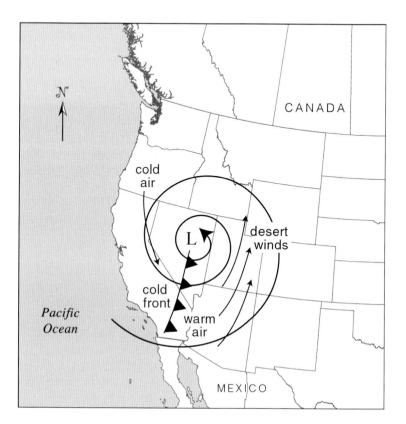

"Desert winds" blow ahead of a cold front moving across Nevada and southern California. The dry, relatively warm, strong southerly winds push unseasonably warm weather to western Utah before the cold front sweeps through, dramatically dropping the temperature and shifting the wind to a northwesterly direction.

of the wind can easily blow lightweight trucks and campers off north-south-oriented roads.

Another dominant wind in Utah is locally called the "desert wind." A strong southerly wind, it blows across the western portion of the state ahead of approaching cold fronts. As cold fronts move eastward across the Great Basin, strong southerly winds develop ahead of the front as a pressure gradient develops between the relatively higher pressure ahead of the storm and the storm's lower pressure. These southerly winds get their name because they originate in the hot, dry deserts of southern Nevada and southern California. The onset of a strong, warm southerly wind in Utah signals that a cold front is approaching.

Desert winds often sweep up loose sand and dirt from the Sevier and Great Salt Lake Deserts, resulting in minor dust storms in the Cedar City–Salt Lake City corridor. The winds also bring about unseasonably warm temperatures and a nice reprieve from winter cold and fog. However, the dusty, short-lived hot spells end abruptly as the cold front and storm system moves across the region. In some cases the warm desert air and cold Pacific storm front clash and create thunder snow, thunderstorms, and very strong erratic winds that shift from a southerly to northwesterly direction. A thunder-snow storm has snow and thunder simultaneously. Thunder snow is relatively rare since cold wintertime surface temperatures prevent strong updrafts of warm surface air from forming; however, with extremely cold air overrunning relatively warm—but still subfreezing—air near the surface, strong updrafts can form. Since the updrafts and downdrafts of a thunderstorm can produce

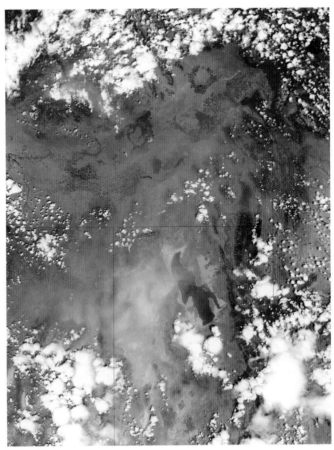

Although typically a wintertime phenomenon, this July 23, 2002, visible satellite image shows southwesterly "desert winds" sweeping across northern Utah. Utah was in its fifth consecutive year of drought, and the winds easily picked up loose soil. Dust appears light brown in this image, whereas fires in southeastern Idaho created the smoke (gray hues) that spans Nevada, southeastern Idaho, and into Wyoming. —Image courtesy of Jeff Schmaltz, MODIS Rapid Response Team at NASA GSFC

condensation and precipitation more efficiently than a uniformly cloudy atmosphere, snowfall rates in thunder-snow situations are astonishing; as much as 5 inches of snow can fall per hour.

Usually, desert winds howl at speeds of 15 to 25 miles per hour, with gusts over 40 miles per hour; however, extreme events brought about by powerful cold fronts sweeping across the Great Basin can whip up winds over 60 miles per hour.

Desert Blast

The strongest desert wind gust ever reported by an official weather station—93 miles per hour—blew at Dugway, Utah, on April 5, 1967.

Desert winds occur year-round but are strongest and most common during the winter. They are a dead giveaway of an approaching Pacific storm and its associated cold front. When the temperature begins to fall and the desert winds shift to a northwesterly direction, residents know the cold front has passed. Once the winds have shifted they are no longer desert winds; they now usher in cold air from the northwest and not warm desert air from the southwest. Usually the post-frontal winds are just as strong (if not stronger) than the desert winds. The highest post-frontal wind gust for a Utah mountain location, recorded just after a cold front had passed, was recorded at Hidden Peak (also known as Snowbird), elevation 11,000 feet, on November 8, 1986. It blew at 124 miles per hour.

Great Basin High and Inversions

Also called "intermountain high" or "plateau high" by meteorologists, the Great Basin high is a dominant weather feature that often establishes itself between the Rocky Mountains and the Sierra Nevada during the winter. This strong and often stationary area of high pressure often centers over Wendover, Utah. Its clockwise rotation weakens and deflects incoming Pacific

What's That Smell?

Most of the time, the smell of decaying brine shrimp—crustaceans of the genus *Artemia* that live in the Great Salt Lake—goes unnoticed along the Wasatch Front since calm winds carry the odor out to the deserts and away from the cities. As the strong winds behind a winter cold front shift from a southerly to a northwesterly direction, however, what the locals call "lake stink" materializes. The pungent odor of dead brine shrimp carries across 80 miles of the Great Salt Lake and into the cities. The potency of the odor depends on the winds across the lake; stronger wind causes more mixing and thus a stronger lake stink.

Luckily for Utah residents, lake stink episodes usually last only a day or so, and in most cases only affect cities north of Nephi. Utah residents know when a cold front has passed by a shift in wind direction and also by the smell of the air.

storms, sending them to the north or south, and the consequences of this weather pattern are felt across the entire West.

Sometimes the high creates a block in the upper-level flow of the atmosphere in the shape of the omega symbol Ω. In meteorological jargon the high is termed an *omega block* or *Rex block*, since Daniel F. Rex first studied it in 1950. These blocks, which form in the usual west-to-east flow of the atmosphere, tend to persist in the same geographic location for several days or weeks. They form most often in winter and spring and are most pronounced near the tropopause (at altitudes of 30,000 to 40,000 feet), but they affect the flow of storms, moisture, and wind all the way down to the ground. The sinking air within the block holds the surface air in place and prevents it from mixing out. This usually means sunny and mild weather with light winds at the ground under the block. While wintertime highs are common, omega blocks are not.

The hallmark of a block is its persistence. It often lasts several weeks, and during this period of time the weather under and to the east and west of the block tends to remain the same. Low-pressure systems—sometimes cut off from the main flow of the atmosphere—on the east and west sides of the block generally keep weather stormy and cool, while tranquil weather persists under the block. Meteorologists still do not know in detail why omega blocks form, and computer forecast models still have trouble predicting their onset or breakdown. Sometimes blocks form when an unusually large ocean storm pumps vast quantities of warm air poleward around the low's eastern periphery.

While an omega block brings long periods of dry weather to western Utah, northern Nevada,

southeastern Oregon, southern Idaho, and northern California, the low-pressure systems pump moisture, clouds, and precipitation into southern California and Texas. Southern California experiences mild, damp weather, while Texas experiences cold, damp conditions. The northern Great Plains, however, can have very cold weather as the omega block drives cold air from the north into this region. Whatever their cause, omega blocks reign as one of the most prominent interruptions of hemispheric circulation. Like a dam in a river, they block the normal west-to-east circulation, which causes the winds around the earth to buckle upstream and downstream of the block. Omega blocks also lead to remarkably persistent weather patterns.

Not all the weather for states under the Great Basin high is good. Unwelcome by-products of this high include stagnant air, inversions, and

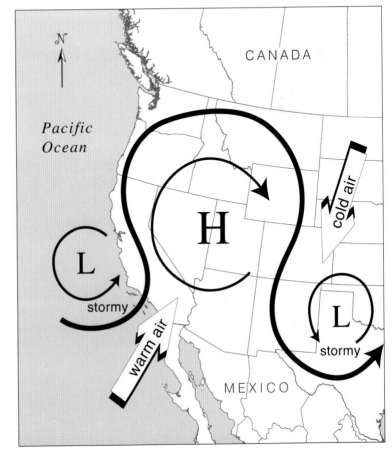

Dynamics of an omega block positioned over the western United States. The circulations around a low-high-low series of systems causes an omega-shaped (Ω) wind pattern in the upper levels of the atmosphere. The sinking air associated with the high pressure brings not only tranquil weather, but also valley fog to Utah, Idaho, and Nevada. Meanwhile, low pressure on either side of the block brings stormy weather; cold temperatures can hit east of the block, while the west side stays mild as warm air blows in from the southwest.

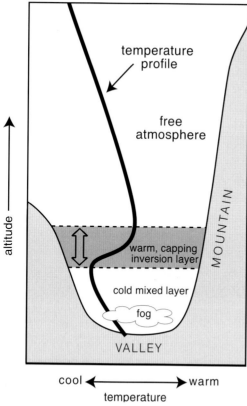

The inversion layer—a tier of relatively warm air—prevents pollutants from escaping into the air above it. Often the colder air beneath the inversion supports fog and low clouds, while areas above the inversion layer experience bright, sunny skies.

cold valley temperatures resulting in lengthy bouts of fog and haze in mountain valleys. The high-pressure area usually produces clear, calm nights, but as the air cools on the mountains and becomes denser than the warmer valley air, it flows downward into the valleys. As a result the coldest air tends to pool in the valley bottoms, driving the temperature near the dew point. This produces fog. Since the morning sun can't penetrate the fog, daytime temperatures in the valleys stay low, and the fog reflects the sun's rays and allows the air over the mountains to heat quickly. The temperature in the mountains can be several degrees warmer than in the valley. This is an inversion.

The term *inversion,* derived from the Latin *inversio,* means "upside down." Typically, as you travel higher into the atmosphere the air grows cooler; in an inversion, however, the air grows warmer. An inversion is nothing more than a warm "lid" on the atmosphere. The sinking air and stable atmosphere prevent the lid from mixing out, plus the relatively calm air under the high pressure allows the inversion to stay intact. Inversions form in stagnant atmospheric conditions, and pollution can become trapped in mountain valleys.

The Salt Lake City–Provo area and the Rocky Mountain Front are especially prone to inversions because the surrounding mountains help trap air. During the notoriously bad winter of 1980–81 a long-lasting inversion spanned across the West, particularly in Utah. Officials reported dense inversion-producing fog at Salt Lake City International Airport for a record 266 hours. Many mountain locations reported warmer mean temperatures than nearby lower-elevation sites. For instance, in January 1981, the average high temperature at the airport (4,222 feet) was 35.8°F, but at nearby Alta, Utah (elevation 8,760 feet), the average high temperature was 39.5°F—3.7°F warmer.

The Rogue River valley in southwestern Oregon has some of the worst inversions in the West. The Medford area, a large valley surrounded by the Cascades and the Coast Ranges, is particularly susceptible to terrible inversions. Between December 16 and 30, 1985, a strong inversion covered the entire Rogue River valley, as well as the Willamette Valley to the north. Dense fog and haze severely disrupted airline travel. The fog-gripped lower elevations achieved much colder temperatures than the higher elevation (generally above 1,500 feet). In fact, the average daily temperature in December at Sexton Summit, Oregon (1,956 feet), was a balmy 44.2°F, 7.4°F above normal. At the same time Medford (1,382 feet) had an average daily temperature of only 36.1°F, 1.6°F below normal. While inversions in Medford and

An inversion over Salt Lake City, Utah. This December 18, 2003, photograph was taken from a helicopter 900 feet off the ground as it entered the foggy, polluted air. The Oquirrh Mountains sit in the distance. —Eldon Griffiths photo

much of the West occur during winter, summertime inversions plague sunny Los Angeles.

Subsidence inversions, produced by compressional heating (the warming of a layer of sinking air), most commonly occur along the immediate coastlines of California. Warm air aloft coupled with cool, surface marine air produce a strong and persistent subsidence or temperature inversion over Los Angeles. The stable, cool marine air near the surface does not rise, allowing warm air above it to continue to warm. This stratification of the lower atmosphere can remain the same for weeks. The cool marine air sometimes creates fog at night. Since the inversion acts as a lid, the pollutants emitted by automobiles, factories, and fires cannot escape and mix into the atmosphere, and "smog" results (the term combines the words *smoke* and *fog*). The Los Angeles inversion occurs 80 to 90 percent of the time between June and October, even though the sun still shines through the haze, light fog, and trapped pollution.

With only a slight gain in elevation, one can escape the polluted, sometimes cold air of an inversion. The higher elevations enjoy not only more sunshine but also considerably more warmth. During a Great Basin high, Alta, Utah, at 8,553 feet, will often see record warmth while the Salt Lake City–Provo valley, in the 4,000-foot elevation range, will shiver with temperatures 20°F colder. If you find yourself trapped in an inversion and you want sunshine, do as you would in a flood—head for higher ground. If driving, slow down and watch for patches of dense fog and black ice that inversions can bring.

State of Extremes: Nevada

The Great Basin region defies simple definition. Hydrologically, it is a place no rivers exit.
Water can only escape through evaporation.
Geologically, it encompasses the Basin and Range
Province, which stretches from Utah's Wasatch
Range to the Sierra Nevada. The Great Basin is
a piece of North America's crust being stretched
by Earth's tectonic activity. Deep basins, which
continue to sink along faults, separate the region's
many north-south-trending mountain ranges.
In general terms, the Great Basin has a desert
environment with mountain summits that reach
12,000 to 13,000 feet and valley elevations that
range from 5,000 to 8,000 feet. A vast area cut off
from the westerly flow of Pacific moisture, its dry
air and high altitude cause some of the country's
greatest daily temperature swings.

Many regions experience extreme temperature swings. But, uniquely, Nevada's huge diurnal
temperature swings take place on relatively calm
days weather-wise, days lacking cold fronts,
downdrafts from thunderstorms, or Chinook-
type winds—weather patterns that lead to large
temperature swings in other regions. The temperature swings in Nevada occur simply through
the processes of solar heating during the day
and, conversely, radiational cooling at night.
Clear skies, low relative humidity, and abundant
sunshine are the key ingredients for quick heating and cooling, and the high desert region of
northern Nevada has them all in abundance.

The very dry air over the Great Basin averages only 37 percent relative humidity during the
summer in Reno. Since dry air has the capability
to warm and cool faster than humid air, the diurnal temperature swings in the Great Basin can
be extreme. The dry air largely results from the
relatively high elevation of the Great Basin—the
decreased air pressure at these higher elevations
causes a decrease in air density, and thus a drop
in humidity. During the day in the desert, the sun
heats the ground, and the ground then warms a

small layer of air above it, creating a transfer of
heat called *conduction*. Once warm enough, the
air rises and mixes with the relatively cooler air
above it, transferring some of its heat—a process

The Driest State

Since the Great Basin lies in the rain
shadow of the Cascades and Sierra Nevada,
it is extremely dry. Much of northern
Nevada and northwestern Utah receive
a meager 5 to 10 inches of precipitation
a year. These mountains force moisture-
laden, eastward-moving Pacific storms
to drop their moisture as they travel up
the windward sides of the mountains.
After crossing the mountain ranges, the
generally weakened storms dry out even
more through adiabatic warming as they
descend the leeward sides of the mountains
and leave the Great Basin parched.

Nevada ranks as the driest state in
the country, with a statewide average
annual precipitation of only 8.68 inches.
Not surprisingly, Reno and Winnemucca
rank among the driest U.S. cities. Reno,
with its average annual precipitation of
only 7.49 inches, is the seventh driest
U.S. city, and Winnemucca ranks as the
tenth with 7.87 inches. Winnemucca and
Reno also rank as the ninth and tenth
least humid cities in the nation, with an
average annual relative humidity of 49
percent and 51 percent, respectively.

In contrast, the highest elevations of
the Ruby Mountains, southeast of Elko, are
some of the wettest areas of Nevada. They
receive upwards of 40 inches of annual
precipitation. The Ruby Mountains, with
many peaks over 10,000 feet, soar high
enough to squeeze moisture out of the
dry air. The wettest area of Nevada, about
15 miles west of Reno at the Mount Rose
Ski Area in the Sierra Nevada, typically sees
about 50 inches of precipitation each year.

termed *convection*. Since the air has low moisture content, the convective process works quickly, unhindered by the evaporative process, which absorbs some of the sun's energy.

At night, with the air no longer heated from below, convective mixing ceases, and with little moisture to trap heat, it dissipates relatively quickly in the atmosphere. Moist air, which contains more water vapor than dry air, can hold heat longer than dry air because it requires more heat energy to heat the water vapor until it evaporates. Without mixing from convection or winds, the air in contact with the ground cools off as the ground itself cools. Since air becomes denser as it cools, it stays near the ground and continues to cool, resulting in a shallow layer of very cold air near the ground beneath a layer of relatively warm

air. This type of temperature gradient, called a *nocturnal inversion,* can cause low evening temperatures in summer.

Northern Nevada is nationally known for its phenomenal, regularly occurring diurnal temperature ranges during the summer months. Unlike areas of the northern Great Plains and northern Rocky Mountains where battling air masses and warm Chinooks often cause great temperature swings in the winter, temperature swings in Nevada occur during tranquil weather. Each summer, residents of northern Nevada feel a diurnal temperature range of 50 to 60°F at least once. On August 27, 1992, the morning low at Elko fell to a chilly 24°F, but by that afternoon the mercury had skyrocketed to a warm 85°F, a range of 61°F. Keep in mind this magnitude of variation rarely happens; however, the temperature often ranges 30 to 40°F every day of summer. The average July high temperature at Elko is 90.4°F, while the average low is 48.6°F—a range of 41.8°F.

To put these temperature ranges in perspective and to illustrate just how remarkable they are, consider the following graph of a typical week during the summer, comparing Reno with San Francisco. The average daily temperature range for San Francisco was only 14.5°F, but it was 47.9°F for Reno. On the thirtieth, the temperature swung 60°F at Reno!

Remarkable Diurnal Temperature Ranges in Northern Nevada

CITY	DATE	HIGH	LOW	RANGE
Elko	August 27, 1992	85°F	24°F	61°F
Reno	August 24, 1970	97°F	37°F	60°F
Elko	August 30, 1995	92°F	32°F	60°F
Reno	August 8, 1969	100°F	41°F	59°F
Reno	September 9, 1971	82°F	23°F	59°F
Elko	August 26, 1992	79°F	20°F	59°F

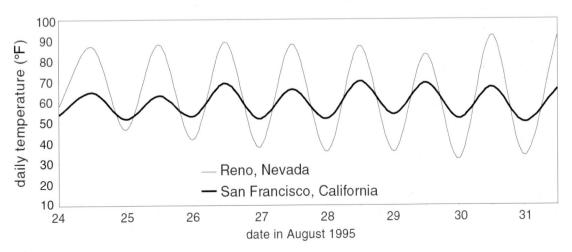

Residents of northern Nevada are not the only ones with the challenging task of choosing what to wear on a summer day. Similar swings in the daily temperature also occur in eastern Oregon, southern Idaho on the Snake River Plain, northern Arizona, and in the valleys of the northern Rocky Mountains. At the other end of the extreme-temperature-swing spectrum, tropical islands near the equator rarely see the temperature vary more than 10°F in the entire year.

The Elko Flash Flood of 1970

Even dry Nevada occasionally receives deluges of rain from tropically driven thunderstorms. In the summer, moist air from Mexico sometimes settles over the Great Basin. As the air warms it rises and cools, forcing water vapor to condense into clouds; eventually strong thunderstorms develop.

On August 27, 1970, a thunderstorm dropped 3.64 inches of rain in one hour in Elko, a record that still stands as Nevada's greatest one-hour rainfall; 4.13 inches fell in two hours. The storm was very localized and confined to an area in and around Elko. The resulting flash flood caused extensive damage to many homes and businesses. The worst damage occurred between Fifth and Sixth Streets in northern Elko, where floodwaters ripped out large chunks of road. The flood caused an estimated $100,000 ($482,000 in 2002 dollars) in property damage. The National Weather Service ranks this remarkable occurrence as Nevada's eighth most significant weather event of the twentieth century.

8

Northern Rocky Mountains

Elevation: 4,000 to 13,000 feet

Natural vegetation: needleleaf evergreen trees

Köppen climate classification: Highland

Most dominant climate feature: cold, snowy winters with cool summers

Snow cover, high elevations, and the sun's low angle in the northern Rocky Mountains make this area particularly cold in January and February. Overnight lows typically range in the −5 to 20°F range, with highs in the teens and 20s. Total precipitation (snow-water equivalent) of 2 to 6 inches with locally higher amounts fall over extreme northern Idaho and western Montana during the two-month period.

Daily temperatures begin warming in March and rise from the 30s and 40s to the 40s and 50s by April. Average morning lows rise from the teens in March to the 20s in April. For many locations, late winter and spring deliver the windiest, wettest, and cloudiest time of year. Total precipitation, which falls mostly as snow, ranges from 1 to 6 inches—an average March-April snowfall of 2 to 8 inches in the valleys and up to 60 inches in the mountains.

Although freezing temperatures and snow are still common in May and June, this period marks the beginning of the mild summer this region generally enjoys. In May, snow begins melting at higher elevations. Average daytime high tempera-

tures warm into the 50s and 60s in May and into the 60s and 70s by June; average morning lows for both months hover in the 30s and 40s, but the higher elevations commonly see pockets of 20s. Precipitation amounts decrease across southern Idaho and northwestern Wyoming, but in general the region receives 1 to 6 inches for May and June combined. With the exception of extreme western Montana and northern Idaho, May and June are the wettest months of the year. The precipitation still falls as snow in the higher elevations, but it usually takes the form of thundershowers at lower elevations by late June.

The warmest and driest months of the year are July and August. Daytime temperatures typically reach into the 70s and 80s while overnight lows can chill to the 30s and 40s. Precipitation totals range from 0 to 4 inches; most of it falls from frequent, short-lived, often fast-moving thunderstorms. Daily thunderstorm activity commonly occurs, especially in the high elevations (generally above 6,000 feet) where mountains provide storm clouds the lift they need to become thunderstorms. Snow is rare although not unheard of in July and August, particularly at the highest elevations (generally above 10,000 feet).

By September cooler temperatures and the first snowfall signal the return of winter weather. By October, snow and much colder temperatures are common, especially in higher terrain. Dramatic temperature and weather changes from one

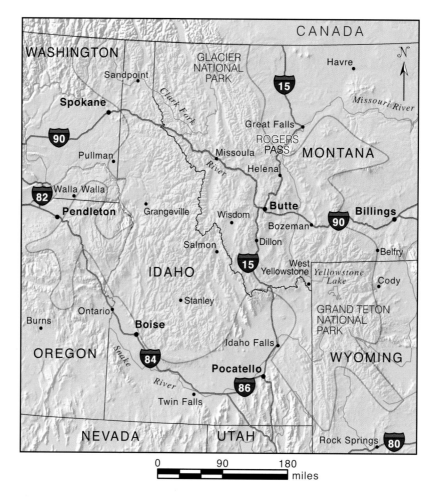

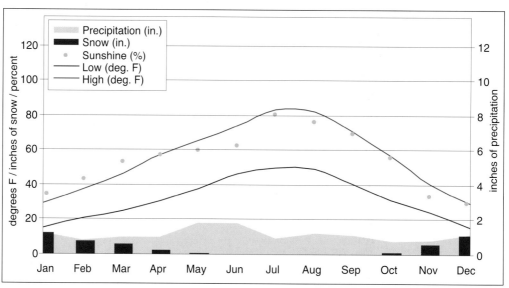

Climate of Missoula, Montana

day to the next commonly occur in October; the warmth of summer radiates one day, and a strong cold front can bring mountain snow the next. Average daily high temperatures plummet from the 70s in September to the 40s and 50s by October. Morning lows drop from the 40s in September to the 20s and 30s in October. Usually September brings only 1 inch or less of precipitation, but 1 to 4 inches fall in October, mostly as snow.

By November and December, snow, wind, and cold temperatures dominate the weather. Average temperatures remain below 30°F, with precipitation totals reaching 10 inches in mountains of extreme western Montana and northern Idaho. Elsewhere, precipitation ranges from 2 to 5 inches, with most of that falling in December. Deep blue skies and brilliant sunshine appear between storms; however, valley locations often suffer from fog and low clouds as a result of inversions. The higher elevations can be hit hard with very heavy snow—snow measured in feet rather than inches—while valleys typically receive 1- to 4-inch snowfall events.

Rime Ice

The northern Rocky Mountains see a variety of icing events because of their closeness to the moisture-rich Pacific Ocean and arctic air masses that move over this region regularly. The mountains become coated with ice, rime ice, hoarfrost, and frost every year. Although the Cascades remove most of the moisture from weaker Pacific storms before they hit the northern Rockies, often enough remains to support rime ice and hoarfrost at the higher elevations of eastern Washington, northeastern Oregon, northern Idaho, western Montana, and Wyoming. Warmer, lower elevations cannot support prolonged periods of rime ice; however, short periods of frost and hoarfrost commonly occur.

Although they form similarly, each type of ice has unique characteristics.

Clear ice: Smooth, compact rime, sometimes called *glaze*. Usually transparent, it has a smooth surface. Factors that favor clear-ice formation include large water droplets and rapid accretion with temperatures at or just below freezing; freezing rain often turns into clear ice.

Rime ice: An opaque white or milky deposit that forms when supercooled water droplets come into contact with an exposed object and freeze rapidly; composed of discrete ice granules visible to the naked eye (unlike clear ice). Rime is denser and harder than hoarfrost, but lighter, softer, and less transparent than clear ice. Small droplet size, slow accretion, and very cold temperatures favor rime formation.

Hoarfrost: Also known as *white frost;* deposit of interlocking ice crystals called *hoar crystals*. Basically large frost crystals; forms when water vapor in a humid air mass comes into contact with stationary subfreezing objects, such as tree branches, plant stems, leaf edges, wires, and poles. The deposition of hoarfrost is similar to the process by which dew forms, except for the below-freezing temperature of the exposed objects.

Frost: A thin, fuzzy layer of ice crystals deposited on an object, such as a window, bridge, or blade of grass. Similar to dew, frost forms when a humid air mass comes into contact with an object that has a temperature at or lower than the dew point of the air mass. Water vapor in the air mass condenses and then freezes on the object.

Often associated with ice that forms on an airplane's wing while in flight, rime ice also occurs at Earth's surface. It forms when supercooled fog or cloud droplets blow onto objects. (The moisture content of a cloud or fog remains liquid even though its temperature is below the freezing point.) When the supercooled moisture hits subfreezing objects, it freezes and accumulates. The

An evergreen tree encased with a thick layer of wind-driven snow and rime ice atop Mount Bachelor in Oregon in 1998.

encased in rime ice, snow, and frost. Sometimes this conglomeration can reach a thickness of several feet.

On January 3, 1961, a three-day-long event produced an 8-inch-thick accumulation of rime ice in Grangeville, Idaho, and established the record rime ice accumulation for the western United States for a single event. A strong area of high pressure centered over Washington and deflected moist Pacific storms north and then southeast of the Pacific Northwest. The high-pressure area caused inversions, which trapped subfreezing moist air near the surface. Heavy fog blanketed much of northern Idaho and produced heavy rime ice accumulations from Grangeville to the Canadian border. Winds blew the fog and freezing drizzle into heavy rime deposits. The heaviest deposits hit power lines, trees, fences, signs, and any other exposed objects and caused widespread damage and power outages.

Clear ice most commonly forms in the valleys of eastern Washington, Idaho, and northern Oregon because temperatures often hover at or near freezing, creating freezing rain. Many valleys average five to ten glazing events per year, with minimal accumulations of less than $\frac{1}{10}$ inch.

The Great Blizzard

On February 1, 1951, the greatest clear-ice storm on record in the United States produced an ice glaze up to 4 inches thick from Texas to Pennsylvania. Twenty-five people were killed in car accidents and by falling trees, and some died of hypothermia. The storm interrupted communications and utilities for a week to ten days across this region and caused five hundred serious injuries and $100 million ($689.7 million in 2002 dollars) in damage.

rapid freezing causes a white or milky granular deposit of ice. Rime ice is often confused with hoarfrost (see list above).

The heaviest accumulations of rime ice occur on mountaintops and ridges where the temperature often stays below freezing and strong winds carry supercooled water droplets. Rime can easily coat the windward sides of trees and buildings with several inches of white ice. On average, the thickest rime ice in the United States occurs at the summit of Mount Washington in New Hampshire (6,288 feet), where accumulations frequently exceed 12 inches. Rime can easily create a wintry-looking landscape. By midwinter, trees in the northern Rockies can become completely

If not for rime ice, several of the Christmases I have enjoyed in northern Idaho and eastern Washington would have lacked a wintry scene.

But thanks to the heavy accumulations of rime ice and hoarfrost, the landscape appeared whiter than it would have if snow had just fallen. Rime ice not only creates a beautiful scene, but also provides an important source of moisture for high-elevation plants. Furthermore, a coating of rime ice provides trees and plants an important insulating blanket, protecting them from cold wintertime temperatures.

Coldest Place in the Lower Forty-Eight: Rogers Pass, Montana

On the morning of January 20, 1954, the weather observer at the Rogers Pass weather station—at 5,470 feet 35 miles north-northwest of Helena—etched a mark on the thermometer where the mercury had stopped. As he did so, he held his breath so it wouldn't warm the thermometer and disturb the extremely low reading. He then sent the thermometer to the National Bureau of Standards for calibration. They confirmed his reading, and the temperature officially became the lowest temperature ever recorded in the contiguous United States: −70°F. (Similar, but not as exhaustive steps are taken today to determine new record temperatures.)

With a hefty 62 inches of fresh snow on the ground and crystal clear skies, rapid radiational cooling coupled with an arctic air mass drove the temperature that low. Although sunny that day, the temperature struggled to reach a daytime high of −22°F, but it shot to a balmy 36°F two days later on January 22. Remarkably, at the moment Rogers Pass shivered at −70°F, Yuma, Arizona—about 1,000 miles away—basked in 55°F weather; a phenomenal 125°F temperature difference across the West.

With nothing but barbed wire fences between Montana and the Northwest Territories of Canada, icy air masses move in effortlessly. Although the air warms slightly during its journey from the north, it sometimes reaches the United States having warmed very little. Since there aren't significant west-east-oriented mountain ranges in its way, the air encounters little resistance, and it warms little while traveling over snow-covered terrain. Arctic air masses can force the temperature to well below zero in Montana for days on end as they undercut the existing warmer air. A light breeze can drop the temperature to −100°F or colder with windchill, making prolonged exposure life-threatening to man and beast. Often these air masses lack the strength to overtop the Continental Divide, so they stall over Montana; however, a strong pressure gradient sometimes causes them to slosh over the divide and fan out across the Snake River Plain (see *Sloshers and Northern Invasions* in *chapter 7*).

Five Western Weather Stations Most Regularly Reporting the National Daily Low Temperature

West Yellowstone, Montana

Stanley, Idaho

Wisdom, Montana

Gunnison, Colorado

Fraser, Colorado

Montana's −70°F reading, well shy of the world record of −127°F set at Vostok, Antarctica, still ranks as one of North America's coldest readings. The coldest all-time temperature recorded in North America is −81°F, recorded at Snag, Yukon, Canada, closely followed by a −80°F reading at Prospect Creek, Alaska—a pipeline camp 25 miles southeast of Bettles along the Dalton Highway.

So how cold can it get on Earth? Meteorologists estimate that during a calm, clear polar night the temperature just above a snowy surface can plunge to less than −140°F. Since weather observers record official air temperatures 5 feet above the surface, meteorologists have not confirmed this. The coldest temperatures occur near the surface since the air remains unmixed with relatively warmer air above.

Idaho's Coldest Spots

On the morning of January 18, 1943, the temperature at Island Park Dam, west of Yellowstone National Park, plunged to −60°F, and this remains the all-time coldest temperature ever officially recorded in Idaho. About 12 miles southeast of the Continental Divide at an elevation of 6,285 feet, Island Park Dam is one of the coldest places in Idaho. The Island Park Reservoir, behind the dam, sits in a caldera, a crater that formed in a volcanic eruption 1.3 million years ago. The caldera has a diameter of 23 miles across at its widest point, and a dense pine forest covers most of it.

Although Island Park Dam has measured the coldest temperature in the state, Stanley, Idaho, along the Salmon River, has colder average wintertime temperatures. Stanley, located in a 6,300-foot-high valley called Stanley Basin, experiences rapid nighttime cooling because it has relatively dry air, resides at a high elevation, and frequently has snow cover. The snowpacked mountainsides experience very strong radiational cooling. The nightly descent of cold, dense air into the valley from high atop the surrounding mountains causes the air temperature at the valley bottom to drop. This air becomes trapped in the basin, driving temperatures to low levels. Freezing temperatures are commonplace throughout the year, with wintertime minimum temperatures often dipping into the 0 to −10°F range. The coldest temperature ever recorded at Stanley was a numbing −54°F on December 23, 1983.

The average wintertime temperature of 3.5°F at Island Park Dam compares to −0.9°F at Stanley. Stanley reports some of the coldest temperatures for the contiguous United States about a dozen times each year, mostly during the summer months. Average summertime (July through August) low temperatures at Stanley dip to the mid-30s, but temperatures have fallen into the midteens. For instance, the temperature at Stanley dropped to 16°F on August 24, 1992.

Precipitation Extremes: Montana

Average annual precipitation varies widely in Montana, and as in many other western states topography governs precipitation. Montana, with a land area of 145,556 square miles, ranks as the fourth largest state in the nation. The far western portion claims the mountains, while the eastern half makes up part of the Great Plains with occasional wide valleys and isolated groups of hills. In general, the western portion of the state receives more precipitation than the eastern part; however, a few valleys in the western portion stay very dry.

The majority of precipitation (53 percent) in Montana falls from afternoon and evening thunderstorms during the summer. The higher elevations, however, receive additional precipitation during the winter as snow—particularly the higher elevations of western Montana. By the time westerly Pacific storms reach the mountains of Idaho and Montana, the Cascades have largely tapped their moisture; however, the storms do have enough water vapor left to support heavy precipitation at higher elevations. The mountains of western Montana and northern Idaho (relative to the desert of eastern Washington) force the water vapor in these weakened storms to rise, cool, and condense into clouds and precipitation. Known as *orographic precipitation*, this is the major reason Montana's mountains receive more precipitation than the valleys during winter. The mountains throughout western Montana place eastern Montana—and some of the western valleys—in a rain shadow, setting the stage for drastic precipitation differences.

The two driest areas of Montana occur in the Clarks Fork Yellowstone River valley in Carbon County and the Jefferson River valley in Madison and Jefferson Counties. Lying in the rain shadow of the mountains of the Absaroka-Beartooth Wilderness, the Clarks Fork Yellowstone River valley receives an average annual precipitation of 8 inches. In fact, in 1960 Belfry, on the river, received only 2.97 inches of precipitation, establishing Montana's record for the least amount of precipitation in one calendar year. The Jefferson River valley, between Whitehall to the north and Dillon to the south, receives an average annual precipitation of 8 to 10 inches since it lies in the rain shadow of the Pioneer Mountains.

Although these spots represent the driest areas of Montana, much of the state receives less than 15 inches of precipitation a year; almost every locale north and east of Great Falls, near the Rocky Mountain Front, receives an average 12 to 15 inches a year. Portions of central and eastern Montana receive the majority of their average annual precipitation in late spring and early fall as deep moisture streaming in from the south wraps around a strong low-pressure system over Wyoming.

The rugged topography of western Montana can produce dramatic ranges in localized precipitation; in some cases, valleys receive five times less precipitation than nearby mountaintops. For instance, the wettest spot in the Bitterroot Mountains, which run along the Montana-Idaho border, sits just west of Stevensville, Montana. It receives an average 80 inches of precipitation (including melted snow) a year, while Stevensville, in the Bitterroot Valley, only receives 12 inches—a difference of 68 inches over 12 miles.

The wettest area in Montana, and in the U.S. Rocky Mountains, is in the northern portion of Glacier National Park, where the average annual precipitation at some locations can reach 120 inches. The wettest areas of Glacier National Park lie west of Saint Mary, Montana, atop Cataract Mountain (10,009 feet), and in the mountains northwest of Summit, Montana. In 1964, Summit, 25 miles southwest of Browning, reported 55.39 inches of precipitation, making that the all-time wettest year ever recorded at a Montana weather station. On average, most of Montana's Rocky Mountains receive 30 to 60 inches of precipitation a year, although the Bitterroot Mountains along the Idaho-Montana border—relatively closer to the Pacific Ocean—can receive up to 90 inches in places.

A large percentage of this region's precipitation falls as snow, which supports some of the largest active glaciers in the conterminous United States. Perched on the Continental Divide at 5,233 feet, Summit, a railroad siding and maintenance station, receives a whopping 241 inches of snow a year, with measurable snow every month except June, July, and August. In 1970, 420.2 inches of snow fell in Summit, establishing the all-time seasonal snowfall record for Montana. Less reliable measurements suggest the snowfall in the surrounding mountains easily exceeds 600 inches every winter. Often harsh winter conditions can sometimes last well into June and sometimes July. On July 7, 1981, a snowstorm with 90-mile-per-hour winds dropped 10 inches of snow on Glacier National Park. On the same day, not too far south in the Rocky Mountains, the temperature reached a record high of 101°F at Denver, Colorado.

Almost half of the annual precipitation in the Rocky Mountains of extreme western Montana and northern Idaho falls between December and February, making it virtually impossible to keep most mountain passes open during the winter. By June snowplows and snowblowers have carved out the roads in Glacier National Park. With snowbanks twice as high as a car in June, a drive through this national park reveals that this is the wettest place in the nation's Rocky Mountains.

A bulldozer removes snow from Going-to-the-Sun Road in Glacier National Park. —Photo courtesy of Glacier National Park

Going-to-the-Sun Road

Built in the early 1930s, this narrow, 52-mile-long road remains one of Glacier National Park's premier attractions. It traverses the park from the west entrance in West Glacier, Montana, to the east entrance in Saint Mary, Montana. The road crosses the Continental Divide at Logan Pass at an elevation of 6,880 feet. Due to inclement weather and heavy snow, the Logan Pass section of the road closes every fall. Spring opening of this and other roads throughout the park typically begins in April with great anticipation by the public and park employees.

The spring clearing of Going-to-the-Sun Road is a significant undertaking, and the date of opening depends on snow conditions atop Flattop Mountain, monitored at a SNOTEL (for SNOwpack TELemetry) station. Approximately 9 miles from Logan Pass, this station is one of six hundred across the western United States. The Natural Resources Conservation Service installs, operates, and maintains the extensive, automated SNOTEL system to collect snowpack and related climatic data. As soon as the Flattop Mountain SNOTEL station indicates consistently melting snow in spring, crews begin digging the road free. This consistent melting typically occurs in mid-April, making the average opening date for Going-to-the-Sun Road June 7.

The Teton-Yellowstone Tornado

July 21, 1987, began uneventfully in the spectacular Grand Teton National Park in northwestern Wyoming—clear skies, cool temperatures, and light winds. By noon, though, puffy cumulus clouds began to thicken over the region, and by 1:00 PM, unusually severe thunderstorms swarmed over Jackson, Wyoming. Thunderstorms aren't unusual in this region, but the unlikely was about to happen.

As the severe thunderstorms traveled in a northeasterly direction, they continued to build steam. Then at 1:50 PM the sky turned greenish and a rare tornado touched down 21 miles northeast of Jackson, near the Box Creek trailhead. The violent tornado grew to nearly 2½ miles wide. Devastating winds tore out, blew down, and snapped 15,000 acres of mature lodgepole pine. Some of the trees, which stood 80 to 100 feet tall, were jerked from the ground and dropped some distance away. The massive blowdown stretched 24.3 miles. From the Box Creek trailhead, which is 10 miles east-northeast of Moran, Wyoming, the tornado ran north beyond the Yellowstone River. Miraculously no one was injured or killed in the incident.

Green Sky

Why the sky often turns greenish just before a tornado remains a mystery. One hypothesis suggests that thick thunderstorm clouds, rain, and hail scatter sunlight in such a way that the clouds take on a blue hue. Then, as the sun gets lower in the sky in the late afternoon or evening when thunderstorms generally occur, the sunlight gets reddish as it travels through Earth's atmosphere and turns the blue hue greenish.

Mountain Tornadoes

Records show only four accounts—other than the Teton-Yellowstone Tornado—of tornadoes occurring at high elevations (greater than 8,000 feet) in the West.

December 2, 1970: A rare December tornado touched down at around 8,000 feet in the Timpanogos Divide area in the Uinta National Forest in Utah. The white tornado tracked across a 38-inch snow cover, spewing snow 1,000 feet into the air, and traveled in a southwesterly direction for about 1 mile. The tornado stretched about ¼ mile wide and toppled trees of 1 foot diameter. Staff of the Timpanogos Cave National Monument reported hearing a loud roaring sound as the tornado dipped down across the divide.

August 7, 1989: A weak twister touched down near Longs Peak, in Rocky Mountain National Park in Colorado. Poor documentation left no other information about this event.

August 11, 1993: A cluster of late-afternoon severe thunderstorms developed across the Bear River Mountain Range, part of which occurs in Duchesne County in Utah. The thunderstorms produced 1-inch-diameter hail, microburst winds, and most remarkably, an F3 tornado at an elevation of 10,800 feet. Uprooted trees damaged four vehicles.

July 12, 1996: Meteorologists documented a tornado and large hail in the Rocky Mountains near Divide, Colorado. The tornado touched down just north of Cedar Mountain Road in the Pike National Forest, uprooting or snapping 80 to 100 acres of spruce, pine, and aspen as it crossed Cedar Mountain Road moving south-southwest. The National Weather Service classified the tornado an F1 with a width of 450 feet and a 0.7-mile-long path.

July 7, 2004: A tornado touched down at 11,975 feet near Rockwell Pass in Sequoia National Park in California.

Salt Lake City Tornado

On August 11, 1999, a strong F2 tornado (113- to 157-mile-per-hour winds) tore a path through the Salt Lake City metropolitan area. The midday tornado touched down west of the Delta Center in downtown Salt Lake City and then ended in the northwestern section of town known as the "Avenues." It lasted about fourteen minutes, had an average width of 300 to 600 feet, and cut a path 4.5 miles long. The storm killed one person, injured more than eighty, damaged five hundred trees, and caused about $170 million in damage. The National Weather Service ranks it as Utah's second most significant weather event of the twentieth century.

This astonishing tornado ranks as the highest elevation F4-tornado ever documented (see F-scale table in *Tornado Alley* in *chapter 11*). When the tornado passed over the Continental Divide just above timberline, it reached its highest elevation—10,072 feet.

High-elevation tornadoes happen infrequently because the supercell thunderstorms that spawn them rarely form in the mountains. Rough, mountainous terrain usually disrupts the complex thunderstorm wind flows—known as wind shear—necessary for supercell development. Not only can complex terrain disrupt the wind shear, it can also disrupt the rotating column of air that makes these thunderstorms so powerful; the variable wind directions of mountainous terrain can also jumble wind shear and the rotating columns of air (see *Tornado Alley* in *chapter 11* for further explanation of tornado development). Supercells need large amounts of warm, humid air rising into

Path of the August 11, 1999, Salt Lake City Tornado —Image courtesy of the National Weather Service Forecast Office, Salt Lake City, Utah

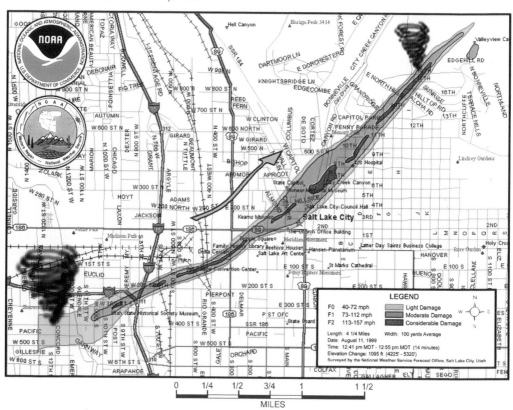

A fire whirl spins out of the Missionary Ridge fire near Durango, Colorado, in June 2002. In the top photograph the fire whirl forms in the fire-heated smoke plume. In the foreground, the winds of the fire whirl blow dust from the parched, drought-stricken bed of Vallecito Lake (elevation 12,010 feet) into its circulation. In the background, smoke and dust rush into the firestorm—against the prevailing wind—as the firestorm pulls air from over a mile away. In the bottom photograph, the smoke-filled fire whirl separates from the plume and travels across the lakebed. A towering cumulus cloud forms above the fire as a result of the smoke plume. Slurry bomber pilots reported that the cloud reached over 30,000 feet in altitude!
—Photos courtesy of Lt. Daniel L. Bender, La Plata County Sheriff's Office

a cold atmosphere to form. This pattern creates the powerful updrafts and downdrafts tornadoes require. In the mountains, air masses tend to be cooler than at lower elevations; hence, even the early stages of tornado development rarely occur at high elevations.

Other tornado-like vortexes caused by major wildfires, not thunderstorms, occur more regularly at higher elevations. Called *fire whirls,* these vortexes occur when superheated air from a fire rises and develops a small low-pressure area. The rotation is amplified by air coming in from different directions, some of which is funneled through canyons or down hillsides. Commonly fire whirls develop on the lee side of ridges where the winds associated with a fire collide with the prevailing winds coming over the ridge. A fire whirl can send smoke, debris, and flames aloft in its vortex, spreading fires, and they can range from less than 1 foot in diameter to over 500 feet.

Perhaps one of the most extreme fire whirls occurred on August 15, 1967, in the Selkirk Mountains of the Idaho Panhandle. A lightning storm ignited a fire near Sundance Mountain, on the southeastern side of Priest Lake, and the fire generated tornado-like eddies that spun over 300 miles per hour. These winds helped spread the fire 20 miles in just twelve hours. The fire burned 50,000 acres and killed two firefighters.

On average, nine tornadoes a year strike Wyoming; over 90 percent of them touch down in the rolling prairies of the southeastern part of the state, where elevations range between 4,000 and 6,000 feet. Wyoming has few tornadoes compared to a place like Texas, which experiences 124 a year. The Teton-Yellowstone Tornado was not only rare for Wyoming, but it was also one of the most unusual ever documented in the United States.

9

Central Rocky Mountains

Elevation: 5,000 to 14,000 feet

Natural vegetation: needleleaf evergreen trees

Köppen climate classification: Highland

Most dominant climate feature: almost daily thunderstorms during summer

Some of the highest peaks in the United States rise in the Central Rocky Mountains (fifty-five in Colorado exceed 14,000 feet), and deep snow cover and dry air make this region particularly cold in January and February. Overnight lows typically fall in the −5 to 20°F range with highs in the 20s and 30s. January and February provide some of the driest months of the year in this region, with 1 to locally 7 inches of precipitation (snow-water equivalent) falling as snow.

By March and April moist spring storms bring slightly wetter conditions, with 2 to 8 inches of precipitation falling mostly as snow. The heaviest snows occur on east-facing slopes during periods of heavy upslope flow associated with storms tracking over Colorado. Daytime temperatures warm from the 40s and 50s in April to the 60s in March, but nighttime lows remain chilly with readings in the 20s and 30s. In early April, the snowpack that began building in October usually reaches its maximum depth, which varies greatly across the region but generally measures between 40 and 90 inches.

With temperatures warming into the 70s and 80s in some locations, the snowmelt peaks in May and June. Low temperatures range in the 30s and 40s. In the valleys the last freezing temperature usually occurs sometime during these months; however, the highest elevations (generally above 8,000 feet) have freezing temperatures year-round. By June most precipitation falls as rain from thunderstorms, but a rare late-season snowstorm can hit.

Thunderstorms in July and August occur like clockwork: they build in the early afternoon, produce areas of severe showers, then quickly dissipate as the sun sets. July is somewhat dry, but August receives a bit more monsoonal moisture (see *Southwest Monsoon* in *chapter 6*). Precipitation for the two-month period varies from ½ inch up to 4 inches. The driest areas occur in Utah's mountains. Average temperatures for both months range in the 50s and 60s. The first freezing temperatures occur in late August.

September is a transition month as monsoonal moisture is no longer available and temperatures begin to cool. By October, snow is fairly common and temperatures continue to fall. Daytime readings drop from the 70s in September to the 60s in October, with overnight lows mostly in the 30s. Dramatic temperature and weather changes from one day to the next are common in October; the warmth of summer can be felt one day, and the next day a strong cold front can bring mountain

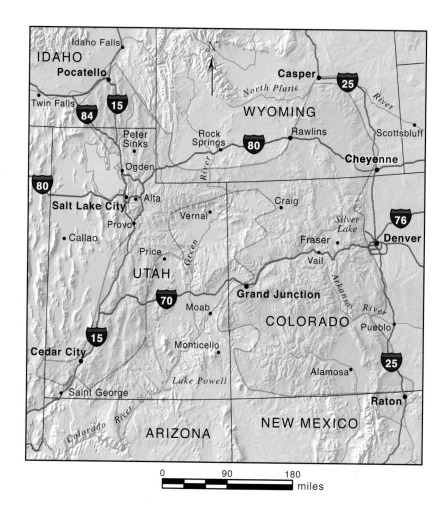

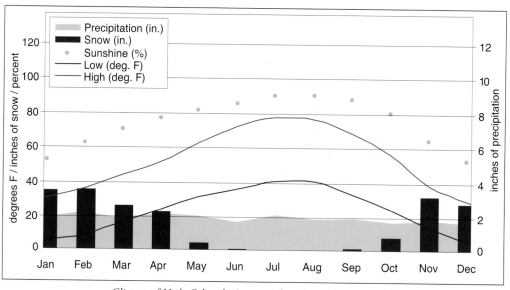

Climate of Vail, Colorado (percent of total sunshine estimated)

snow. Although September is rather dry, with up to 2 inches of precipitation falling, October can bring ½ to 4 inches of precipitation.

By November and December, snow, wind, and cold temperatures dominate the weather. Average temperatures remain below 30°F (sometimes 20°F), and precipitation totals (including melted snow) reach 8 inches in Utah's mountains. Elsewhere precipitation is much lighter; 1 to 5 inches is more common. December is rather dry, with only light fluffy snow making up its total precipitation. Clear skies and dry air are common with cold nights and mild days; however, inversions are also common in the low-elevation valleys. Very strong snowstorms can hit the region, sometimes dropping several feet in one storm.

Champagne Powder and World-Record Snow Intensity: Colorado

Colorado has a well-deserved reputation as one of the best places in the world to ski. The mountains of Colorado receive light, fluffy snow, the foremost reason for the area's superb skiing. Nicknamed the "fluffy stuff," "white gold," or, most popularly, "champagne powder," this light type of snow is what many skiers prefer to carve their turns on. Unlike snow in the Cascades, Alps, or Sierra Nevada, Colorado's snow contains much less water than snow in the areas closer to coasts and creates powder so light you can remove half a foot from your car by blowing on it. The ski areas nearer the coasts may boast massive quantities of snow, with mean annual totals twice that of Colorado, but its powdery quality makes Colorado snow so popular, not its quantity.

The average meteorological conversion of snow to liquid water is 10 inches of snow to 1 inch of water; however, 10 inches of Colorado snow usually contains ⅓ inch of water. In extreme cases, it takes 40 inches of Colorado snow to melt to even 1 inch of water. The fluffiest, lowest-density snow in the central Rocky Mountains typically falls with light winds and surface temperatures near 15°F. Dendrites, the largest of the six types of snow crystals, make up Colorado's fluffy snow.

Snow forms in storm clouds several thousand feet up where cold temperatures rule. Dendrites form when the air temperature in the clouds ranges between −4 and 13°F. Snow crystal shape depends mostly on air temperature, although humidity also plays a role. The ideal surface temperature for champagne powder, between 15 and 25°F, guarantees the snowflakes will retain their crystal structure, which prevents the flakes from packing together tightly. This allows for a lot of air pockets in the resulting snowpack.

Colorado's high elevation and intercontinental location give it the optimum mix of dry, cold air, temperature, and humidity that produces low-density, dry, fluffy snow. Colorado's snow not only depends on a certain air temperature, but on the appropriate relative humidity as well. All snowflakes form in cold air that reaches 100 percent relative humidity, when water vapor in an air mass can no longer remain in its gaseous state. In cold, supersaturated (greater than 100 percent relative humidity) conditions, water vapor changes directly to ice, skipping the liquid phase of water.

Since the quantity of water vapor air can hold depends on its temperature, even the dry air over Colorado can become supersaturated if it is cooled enough. For instance, air at the freezing point needs only 4.8 grams of water per cubic meter to become saturated; any more moisture would make it supersaturated. Compare this to warm (68°F) air, which needs 17.3 grams of water per cubic meter to reach saturation. The vapor-to-crystal transformation begins when supercooled vapor collides with tiny particles of salt, dust, or other ice crystals. The continued cooling or continued addition of moisture beyond 100 percent relative humidity is critical in order for snow to continue forming, otherwise there is not sufficient water vapor to form snow.

The Six Major Types of Snowflake

Star crystals: Form at temperatures near 5°F. One of the most common types of snowflake; several stars may join to form a larger snowflake under ideal conditions.

Dendrites: Three-dimensional star crystals with branches growing from more than one plane. Branches (or arms) connect randomly to a central structure. These complex critters form in extremely cold conditions (–4 to –13°F) in the presence of high levels of atmospheric moisture.

Column capped with plates: Composite flakes formed when a particle of snow passes through different temperature and moisture zones on its journey to the ground. The columns form first, usually at higher and dryer regions of a cloud, and combine with star crystals as they fall through lower and wetter cloud elevations.

Plates: "Wannabe" star crystals lacking moisture; form at temperatures between 14 and –4°F with insufficient atmospheric water vapor available to shape the delicate arms of a classic star.

Columns: Denser, smaller snowflakes that form in dry air and in a wide range of temperatures (5 to –13°F).

Needles: Long and narrow; form at the upper end of the temperature spectrum, usually with ground temperatures at or near freezing. To grow, these crystals need an air temperature in the 14 to 23°F range. Needles tend to produce a dense, stiff snowpack, which can lead to an avalanche under the right conditions.

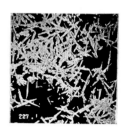

—Wilson Bentley photos, courtesy of the National Oceanic and Atmospheric Administration/Department of Commerce

Official One-Day Record Snowfalls for Western States

PLACE	SNOWFALL (inches)	DATE
Arizona, Heber Ranger Station	38	14 Dec. 1967
California, Giant Forest	60	18–19 Jan. 1933
Colorado, Silver Lake	76	14–15 April 1921
Montana, Summit	44	29 Jan. 1972
New Mexico, Red River	36	12 March 1978*
Oregon, Crater Lake National Park	37	17 Jan. 1951*
Utah, Alta	38	1–2 Dec. 1982
Washington, Winthrop	52	21 Jan. 1935
Wyoming, Burgess Junction	38	14 March 1973

* Record also documented at an earlier date, at the same time, or at another place.

Source: Western Regional Climate Center, Reno, Nevada

Idaho Record Snowfall

On February 11, 1959, Sun Valley received 38 inches of snow in just twenty-four hours, setting the twenty-four-hour snowfall record for Idaho. Weather observers reported 182 inches of snow from the same storm on the ground at Mullan Pass in northern Idaho.

Montana Snowfall Records

The greatest single twenty-four-hour snowfall in the state of Montana is 44 inches, measured at Summit at the southeastern end of Glacier National Park on January 20, 1972. The greatest single-storm snowfall total in the state of Montana is 57 inches, also recorded at Summit, between January 19 and 21, 1972.

The average annual snowfall for Colorado ski areas ranges from a high of 465 inches near Wolf Creek Pass in the San Juan Mountains to a respectable low of 200 inches atop Buttermilk Mountain near Aspen, Colorado. Most ski areas in Colorado open on or slightly before Thanksgiving weekend and close in April, making for a long, enjoyable season.

On the afternoon of April 14, 1921, the first snowflakes of a world-record snowstorm began to fall on Silver Lake in Colorado, 17 miles west of Boulder. Not until about 6 PM the next day did the weather observer attempt to measure the remarkable snowfall. He measured 87 inches of snow that had arrived since the snow began falling at about 2:30 PM the day before; the dry snow contained only 5.60 inches of water. By the time the remarkable 32½-hour storm ended, Silver Lake had received 95 inches of snow (the equivalent of 6.40 inches of water). At the Fry's Ranch weather station, at 7,500 feet in the northwestern portion of Larimer County—about 50 miles north-northwest of Silver Lake—62 inches fell in 22 hours.

Since the Silver Lake observer only measured the snow once in a 27.5-hour period, climatologists had to determine a twenty-four-hour amount for the record books. The U.S. Weather Bureau (currently the National Weather Service) prorated the 87-inch amount to 76 inches and arrived at the official twenty-four-hour snowfall total. And so it was, and still is today, the world's greatest twenty-four-hour snowfall. The Silver Lake record shattered the old record of 60 inches recorded at Giant Forest, California, on January 18–19, 1933. Since then, other states, including Washington, have had unofficial reports of snowfalls exceeding 60 inches in twenty-four hours, but anything over 60 inches is very rare.

A number of ideal atmospheric conditions combined to create the extreme Silver Lake "upslope" snowstorm—a snowstorm fed by a flow of moist air traveling up the slopes of mountains.

A surface low-pressure area was passing slowly across the southern tier of Colorado, rotating low-level moisture from the Gulf of Mexico northward into Colorado (see *Colorado Low* in *chapter 11*). At the same time a large polar air mass dropped south out of Canada and destabilized the atmosphere, causing the relatively warmer air near the surface to rise rapidly into the cold air aloft. Furthermore, a rapidly developing and very strong upper-level low-pressure area over southwestern Colorado provided strong easterly winds and energy. The cold air from Canada and the low-level moisture from the Gulf mixed over the Front Range, and the strong counterclockwise flow of the surface low hurled it all at the Front Range and adjacent foothills as a potent easterly "upslope" snowstorm.

Unlike most upslope storms, which generally leave the inner Rocky Mountains with less moisture than the Front Range, this storm provided the mountains with a lot more snow than usual. Dillon, Fraser, and Grand Lake, in the mountains behind the Front Range, reported some of the largest one-day precipitation amounts for the twentieth century. Oddly, Boulder received a mere 9 inches of snow, although 12 to 24 inches was a more representative total for towns along the Front Range. Depending on how deep and how far off the ground the low-level moisture is during an upslope storm, sometimes the Front Range receives less snow than the inner Rockies. In the April 1921 storm, the moisture was slightly elevated and therefore left the Front Range with considerably less snow. According to the April 1921 *Climatological Data*, snow drifted "to depths of 7 feet in many parts of Denver and 8 feet near Corona, Colorado. Many trees broken and telephone and telegraph poles splintered."

Climatologists thought the Silver Lake snowfall record was broken during an unprecedented lake-enhanced snowstorm at New York State's Tug Hill Plateau, but weather observers there used nonconforming measurement techniques to arrive at their figure. Montague, New York, reportedly received 77 inches of snow in twenty-four hours between January 11 and 12, 1997. After very close scrutiny, federal and state climatologists judged the measurement inaccurate since it was composed of many measurements taken at short time intervals. It was a huge snowfall, but climatologists recalculated the total as 51 inches in twenty-four hours, much less than the Silver Lake storm. Daily snowfall is defined as the accumulation of new snow since the previous day's measurement. For purposes of climatological comparison, daily snowfall should be measured once in a twenty-four-hour period. Since snow settles as it falls, the sum of frequent measurements can misrepresent and skew snowfall statistics. Until the Silver Lake 76-inch amount is exceeded by an observation that conforms to climatological standards, this observation will continue to be acknowledged as the world's twenty-four-hour snowfall record.

Greatest U.S. Snowfall Records

Greatest snowfall for a single storm: 189 inches; February 13–19, 1959, Mount Shasta Ski Bowl, California

Second greatest snowfall for a single storm: 149 inches; January 11–16, 1952, Tahoe, California

Greatest measured snow depth: 454 inches; March 1911, Tamarack, California

World's greatest twelve-month snowfall: 1,140 inches; 1998–99, Mount Baker Ski Area, Washington

Greatest twelve-month snowfall in Alaska: 975 inches; 1952–53, Thompson Pass, Alaska

Greatest twelve-month snowfall in the eastern United States: 565 inches; 1968–69, Mount Washington, New Hampshire

Not Quite the Icebox of the Nation: Fraser, Colorado

Residents of this town, 6 miles northwest of Winter Park, have long claimed Fraser as the "Icebox of the Nation," but climatologists agree that it is just one of many famous "iceboxes" in the lower forty-eight. (There are colder locations in Alaska, so "nation" here refers to the contiguous United States.) Fraser earned its nickname by frequently recording the nation's extreme daily low temperature. For instance, Fraser topped the list of cold spots in 1996, recording thirty-eight of the national daily low temperatures, and it came in second in 1997 with twenty-one.

Like many other cold spots in the Rocky Mountains (such as Wisdom, Montana; Stanley, Idaho; West Yellowstone, Montana; and Gunnison, Colorado), Fraser, near the Continental Divide, sits in a high alpine valley surrounded by mountains. Nestled at 8,568 feet, Fraser is far from moisture sources such as the Pacific Ocean or Gulf of Mexico. With little moisture in the atmosphere to form clouds and trap heat at the surface, Fraser experiences rapid radiational cooling at night. Besides the lack of temperature-regulating moisture in the atmosphere, cold, dense air accumulates—through strong radiational cooling—on the slopes of the 10,000- to 12,000-foot mountains that surround Fraser. This cold air drains into the Fraser Valley at night, helping push temperatures to their extreme lows. Broad valleys, such as the Fraser Valley, are more conducive to extremely cold temperatures than steep valleys and canyons. Cold air that drains into a steep valley tends to accelerate and mix more thoroughly with warmer air above the surface; this results in warmer temperatures. In broad valleys, however, cold air drains more slowly and mixes less. It tends to pool more.

Fraser's notoriety as an icebox comes largely from its extremely cold nights during the summer. The average low temperatures for June, July, and August are 28.8°F, 32.5°F, and 32°F respectively. These temperatures rank considerably lower than those of Gunnison, Colorado—another renowned icebox. Gunnison reports average lows of 36.1°F, 42°F, and 40.8°F for June, July, and August, respectively. Very rarely does Fraser have the coldest average wintertime temperatures. Since Fraser's winter lows are usually offset by rapid daytime heating, the average winter temperatures are somewhat warmer than other cold spots in northwestern Wyoming, northeastern North Dakota, and northern Minnesota.

The all-time coldest temperature ever recorded in Fraser, Colorado, was –53°F on January 10, 1962, just 8°F shy of the state record set at Maybell. On June 25, 1914, the temperature plummeted to 12°F at Fraser, making that the coldest summertime reading in its ninety-year record.

Contrary to what residents may claim, Fraser, Colorado, is not—at least officially—the Icebox of the Nation, but it certainly can call itself the "Summer Icebox of Colorado." International Falls, Minnesota, is widely accepted as the true Icebox of the Nation, although statistics show that several towns have colder average temperatures during the winter months. Langdon, North Dakota, and Warroad, Minnesota, share the unofficial title. However, if you base the statistics on the mean annual temperature, perhaps Fraser, Colorado, deserves it. The average annual temperature at Fraser hovers at a chilly 33.1°F, compared to 36.4°F at International Falls, Minnesota. Truthfully, the "Icebox" title is not based on statistics.

Some consider Minnesota's winter cold a claim to fame. International Falls on the Canadian border copyrighted itself as the Icebox of the Nation to attract not just tourists, but companies wanting to test mechanical equipment in cold weather. This may seem an infamous distinction; however, those who must endure such harsh winters can take some solace in the uniqueness of their weather.

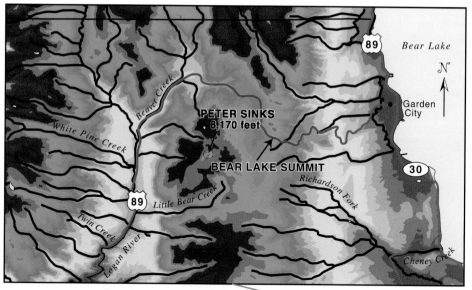

At 8,170 feet, Peter Sinks, Utah, sits in a bowl-like valley surrounded by mountains over 8,800 feet tall. This terrain is ideal for extremely cold temperatures.

elevation (feet)

■ 5,680–5,800	□ 6,601–6,800	■ 7,601–7,800	■ 8,601–8,800
■ 5,801–6,000	□ 6,801–7,000	■ 7,801–8,000	■ 8,801–9,000
■ 6,001–6,200	□ 7,001–7,200	■ 8,001–8,200	■ 9,001–9,200
■ 6,201–6,400	■ 7,201–7,400	■ 8,201–8,400	■ 9,201–9,400
■ 6,401–6,600	■ 7,401–7,600	■ 8,401–8,600	■ 9,401–9,600
			□ 9,601–9,800

UTAH

0   2   4
miles

Icebox of the Nation Title Contenders

LOCATION	AVERAGE WINTER TEMPERATURE 1971–2000 (December–February)
Langdon, North Dakota	4.8°F
Tower, Minnesota	5.5°F
Warroad, Minnesota	5.8°F
Pembina, North Dakota	5.9°F
Hallock, Minnesota	6.0°F
Belcourt, North Dakota	6.2°F
Cavalier, North Dakota	7.1°F
International Falls, Minnesota	7.3°F
Crookston, Minnesota	10.3°F
Fraser, Colorado	14.6°F
West Yellowstone, Montana	17.6°F

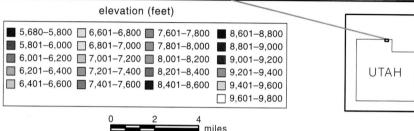

Almost the Record

In 1985, the state of Utah experienced its all-time maximum and minimum temperature extremes—the only state in which both extremes occurred in the same year. For over seventy years Strawberry Tunnel, along U.S. 40 southeast of Heber City, held the Utah record low of –50°F, set on January 5, 1913. But on February 1, 1985, the temperature plunged to –69°F at Peter Sinks, Utah, just one degree shy of the conterminous United States record of –70°F set at Rogers Pass in Montana. Five months later on July 5, Saint George, Utah, sizzled to the state's highest reading of 117°F.

Peter Sinks is located a few miles southwest of U.S. 89 near the Bear Lake Summit in the northeastern part of the state. As the name infers, the topography of the region is shaped like a sink—only water doesn't drain out of this one. The bottom elevation of the sink is about 8,170 feet, and the general rim elevation around it ranges between 8,310 and 8,380 feet, so the depth of the sink is roughly 140 to 200 feet. Peter Sinks, like other nearby sinks, is relatively small, about 3,000 feet wide, and resembles a shallow saucer more than a deep bowl. The air in the bottom of these sinks can be considerably colder—tens of degrees cooler—than just 20 or 30 yards up the sides of the sink. This kind of topographic feature combined with optimum meteorological conditions—an arctic air mass, strong high pressure, clear skies, abundant snow cover, cold air aloft, and low humidity—makes areas like this very susceptible to extremely cold nighttime temperatures. All of these conditions come together every year in Peter Sinks, driving the temperatures to –55°F or colder.

Although the –69°F reading at Peter Sinks is accurate, it is a bit controversial. It was registered during a four-month-long test of a new, highly sensitive, and accurate, monitoring instrument. A thermocouple measures temperature by detecting the change in electrical current through two dissimilar metals. If nearby states had had similar equipment at strategically cold spots, perhaps they too would have established new state records. (Colorado did establish a new record when Maybell dropped to –61°F on February 1.) Without the new equipment, the record –69°F temperature would not exist since the nearest National Weather Service site—Randolph, some 30 miles to the east—reported a low temperature of only –43°F on February 1, the coldest temperature ever reported there. So the controversy is not that the temperature truly bottomed out at –69°F, but rather that other states did not have the same luxury of having thermocouples in remote, cold sinks during the 1985 arctic outbreak.

Most likely, the West has felt even colder temperatures than those reported. A number of small-scale sinks, such as Peter Sinks, dot the West, but we know little about them and, more importantly, they lack thermometers to measure temperatures.

Lake Effect Snow:
The Great Salt Lake

Areas nestled in the northern Wasatch Mountains of Utah receive the most precipitation of any Utah location. The higher peaks—from above 8,000 feet to some over 12,000 feet in a swath just east of Provo extending to northeast of Logan—collect over 65 inches of precipitation a year. In 1983 Alta, Utah, received a phenomenal state-record 108.54 inches of precipitation during the calendar year—far more than any other central Rocky Mountain location receives in an average year. During a normal year Alta picks up 58.45 inches of precipitation, which makes it the wettest town in Utah. Most of the precipitation falls as snow. Many areas easily receive over 200 inches (17 feet) of snow a year, with areas around Alta receiving over 500 inches (42 feet). These are phenomenal amounts of rain and snow considering this region sits next to the Great Basin Desert.

After being disorganized by the Cascades and Sierra Nevada, Pacific storms reorganize while traveling across the Great Basin. Similar to how storms reorganize after crossing the Continental Divide in Colorado, storms crossing the Great Basin undergo intensification as they travel east. As a storm moves over the high peaks of the Sierra Nevada and Cascades, it is squeezed into a thin layer of the atmosphere, bound by the mountains underneath and the top of the atmosphere, which a storm can't penetrate. This forces a storm to spread out horizontally, slowing its circulation and weakening the storm; however, as the storm then moves over the Great Basin, it once again expands vertically given the basin's lower surface elevation and the greater distance between the earth's surface and upper atmosphere. Just as an ice skater pulls in her arms to gain speed, this vertical tightening of the storm system causes its circulation to spin much faster, and the storm regains its strength. Pacific storms are more apt to move across the Great Basin between February and May and in November.

The Great Salt Lake is the remnant of a much larger Ice Age lake called Lake Bonneville, which was so large that its waters lapped at the Wasatch Range's flanks. Today, the Great Salt Lake is much smaller, particularly shallow, and salty. The lake's high salinity prevents it from freezing in winter; consequently, the atmosphere to the south and east—downwind—of the lake has a year-round body of water at its disposal from which to wick moisture. This unique source of moisture, coupled with passing storms, brings lake effect and lake-enhanced snowstorms to the region.

During summer, 90 to 100°F days heat the lake's shallow water to a relatively warm temperature. The high salinity of the Great Salt Lake prevents it from freezing in winter, resulting in warm water temperatures and relatively warm air temperatures just above the lake's surface. Early in winter, with the lake's water still warm compared to the atmosphere, cold northwesterly winds associated with passing storms blow across the lake, creating an unstable atmosphere; the colder the air, the more unstable the atmosphere. Instability promotes uplift and evaporation. A low-level northwesterly wind then entrains this rising, moist air. Eventually, the low-level moisture

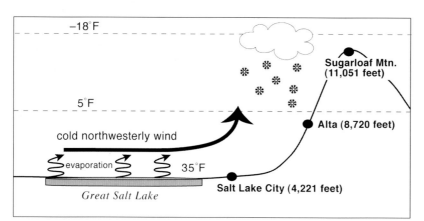

Optimal conditions for lake effect snow in the Salt Lake City area. Cold winds blowing over the unfrozen, open waters of the Great Salt Lake collect moisture. The resulting flow of cold, moist air is forced up the Wasatch Mountains, causing snow-producing clouds.

Utah and Wyoming Rain Shadows

During the 1952 water year (amount of precipitation measured between October 1, 1952, and September 30, 1953) the small community of Callao, Utah, received a mere 0.71 inches of precipitation. This stands as the driest water year ever recorded at a Utah location. Ordinarily, Callao, about 64 miles south of Wendover, receives 6.28 inches of precipitation during a water year, making it one of the driest areas of Utah.

Most of the terrain west of Interstate 15 and south of Interstate 80 in Utah falls in the rain shadow of the Cascades and Sierra Nevada and receives less than 12 inches of precipitation a year; the exceptions are the higher mountains (generally above 8,000 feet), where precipitation amounts range up to 30 inches. Most of the Great Salt Lake and Sevier Deserts receive less than 8 inches a year. Likewise, parts of southeastern and eastern Utah receive little precipitation. Myton, Utah, 40 miles west of Vernal, Utah, reported the driest calendar year ever at an eastern Utah location, with only 1.34 inches of precipitation in 1974. This area of Utah lies in the rain shadow of the Wasatch and various other north-south-trending mountain ranges, such as the Pahvant, Tushar, and Fishlake Ranges that run through central Utah.

Wyoming's driest areas—the Great Divide Basin and the Red Desert—lie just north of the Utah border in southwestern Wyoming. The towering Wyoming Range and the even higher Bear River Range in northern Utah and southeastern Idaho leave the Great Divide Basin and the Red Desert between Rock Springs and Rawlins in a rain shadow. Only 10 to locally 5 inches of precipitation falls in this area a year. The driest year ever at a Wyoming location occurred at Lysite, 115 miles north-northwest of Rawlins, where only 1.28 inches of precipitation fell in 1960.

reaches the end of the lake and blows across much cooler and increasingly higher terrain, from the lake's 4,200-foot level up the 8,000- to 12,000-foot Wasatch Mountains. The air cools and expands (orographic lifting and cooling) as it rises, releasing its hefty load of water in the form of heavy snow.

A lake effect storm can pummel areas east and southeast of the lake with heavy snow, while nearby areas receive only a dusting. More often than not, the Big Cottonwood, Little Cottonwood, and Parleys Canyons, located east of Salt Lake City, frequently clog with several feet of snow, while areas north and south see only light snow or flurries. Sometimes these lake effect storms are so localized that clear skies prevail everywhere except downwind of the lake. For example, on October 17–18, 1984, a storm dumped 27 inches of snow on Holladay, while Granger—only 10 miles to the west—reported only 1 inch.

An avalanche crashing down a steep cliff in Utah's Wasatch Mountains —Photo courtesy of the Special Collections Department, J. Willard Marriott Library, University of Utah

Utah's Coldest Winter

Utah's most severe winter since 1899 hit during the winter of 1948–49. The coldest winter on record, snow fell along the Wasatch Front and nearby mountains in record amounts. Downtown Salt Lake City received 93.1 inches of snow, with 39.1 inches of that falling in December alone; normal winter snowfall reaches 56.3 inches. Ten people died and cold temperatures killed fruit trees and livestock. The average daily low temperature during the winter at Salt Lake City International Airport reached only 10°F—the lowest since records began in 1923—compared to the average wintertime low temperature of 23°F. The National Weather Service ranks this severe winter as Utah's fourth most significant weather event of the twentieth century.

Satellite image of the Great Salt Lake in northern Utah on February 8, 2001. The Great Salt Lake adds moisture to winter storms, causing heavier snowfall east and south (downwind) of the lake. Areas west and farther away from the lake have bare ground. —Image courtesy of NASA/ GSFC/LaRC/JPL, MISR Team

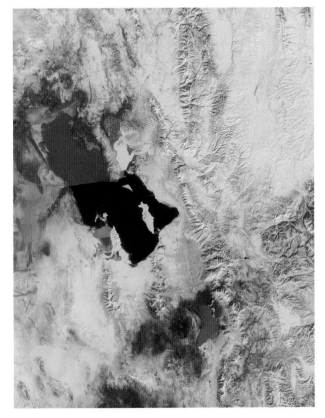

Lake effect snowstorms typically occur during fall and early winter and last anywhere from a few to thirty-six hours but average twenty hours. A number of times each autumn, lake effect storms drop enough snow that officials have to close roads in the Wasatch canyons because of avalanche danger and sometimes because snowplows can't keep up with the intense snowfall. Often skiers and visitors are trapped at ski resorts for days. Later in winter, as the lake's water cools, the winds draw less moisture from the lake. The difference in temperature between the lake and the air traveling over its surface provides the most important factor for a strong lake effect snowstorm.

Lake effect snowstorms typically occur after an easterly Pacific storm and its associated cold front have moved across northern Utah and into eastern Utah. As a storm system passes through Utah with its own instability, uplift, and moisture, it creates its own snow. It also picks up moisture from the lake, which it drops as lake-enhanced snow. Lake-enhanced snowstorms often transition into lake effect snowstorms as a storm and its associated moisture move east of Utah. This weather pattern sets the stage for northwesterly winds, which usher in much colder air from Canada. The primary difference between lake-enhanced and lake effect snow is the extent of snowfall: lake-enhanced events often drop snow over a much larger area since they have more moisture to work with. Lake effect storms, on the other hand, tend to be localized.

Once a storm enters eastern Utah with its moisture, the wind direction shifts, and the snowfall associated with the storm's dynamics ends; but areas downwind of the lake continue receiving snow, only now the Great Salt Lake provides the primary moisture source. Since the longest portion of the Great Salt Lake is oriented from northwest to southeast, the cold northwesterly

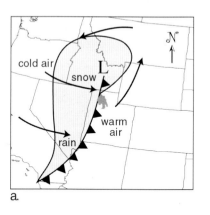

a.

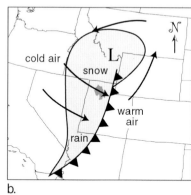

b.

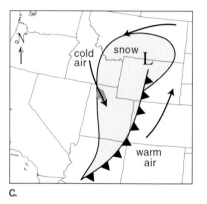

c.

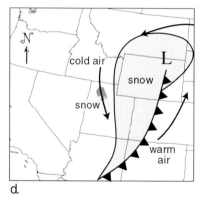

d.

Progression of a cold front across Utah and its interaction with the Great Salt Lake. (a) A cold front approaches Utah with warm air blowing ahead of it, raising the lake's water temperature and increasing evaporation rates. (b) The cold front passes over the lake, entrains moisture, and drops the lake's surface temperatures below freezing. Snow associated with the storm's and the lake's moisture, now called lake-enhanced snow, begins falling in the Salt Lake City region as the air mass hits the Wasatch Range. The heaviest snow falls to the east of the city in the mountains. (c) As the cold front and low-pressure area move east, winds shift to a northwesterly direction, bringing the heaviest lake-enhanced snow to the southeastern portions of Salt Lake City. The snow begins to transition from lake enhanced to lake effect. (d) The storm and its associated moisture have moved out of the Salt Lake City area and the winds have turned north to northwesterly. Lake effect snow bands develop as northwesterly winds blow across the lake, causing heavy accumulations in southwestern and western portions of Salt Lake City but drop little snow elsewhere.

wind has a long fetch of water to blow across. Meteorologists call the northwesterly winds *post-frontal winds* since they occur after a storm's cold front has passed. They call the warm southerly winds that blow ahead of the front *pre-frontal winds,* or as locals call them, "desert winds." These southerly winds blowing off Utah's deserts further warm the Great Salt Lake's surface temperature just before the cold front moves through. These winds are important since they help create an even greater temperature contrast between the lake and the coming cold air.

Lake effect snow often flows to areas east and southeast of the Great Salt Lake in elongated bands of snow-delivering clouds called *lake effect snow bands,* which create some of the region's worst winter storms. Lake effect snow bands tend to develop parallel to the prevailing wind flow in the lower atmosphere, and they typically occur as a prominent single band or an area made up of smaller bands. Single bands tend to exhibit greater intensity and heavier snowfall over a narrow area, while multiple bands deliver lighter snowfall over a larger area. Slight differences in the wind direction and weather pattern can change the snow band location and behavior dramatically.

As the distant storm system continues tracking east, it causes winds across northern Utah to gradually shift from northwesterly to northerly. The lake effect snow bands then take aim at the southern shores of the Great Salt Lake, typically near Tooele. Eventually the storm system moves east over the Great Plains, causing winds to weaken and become variable across northern Utah. This ends the lake effect snowstorm.

The lake's moisture and surrounding topography create large contrasts in average annual snowfall throughout the Salt Lake City region. Salt Lake City International Airport, about 5 miles west of the Wasatch Range, receives approximately 65 inches a year, while the "benches"—an area of Salt Lake City nestled along the foothills of the Wasatch Range—receives nearly 100 inches.

Alta, at 8,750 feet and southeast of Salt Lake City along the Wasatch Range, gets buried by 518 inches of snow a year. In contrast, on the northwestern side of the lake, at the Lakeside, Utah, weather station, the average annual snowfall reaches a mere 19.8 inches.

Moisture picked up from the Great Salt Lake causes significantly more snow to fall in the Wasatch Range than would ordinarily fall near a desert region such as the Great Basin. While it provides the largest inland moisture supply for the West, this pales in comparison to the amount of moisture the Pacific Ocean and Gulf of Mexico storms provide the West. The Great Salt Lake may provide only .5 percent of the West's total annual precipitation.

An El Niño–Induced Blizzard

Many lake-enhanced and/or lake effect snowstorms have struck the Salt Lake City region, but a February 21–27, 1998, event stands out. An unusually warm El Niño winter kept the Great Salt Lake warm enough to produce a major lake effect snowstorm late in winter. The two-day blizzard dropped snow at the rate of 2 to 3 inches per hour, leaving the city under a record-breaking 18 to 28 inches of snow; Bountiful reported 51 inches of snow while Alta reported 87 inches. This went down as the fourth snowiest storm ever in Salt Lake City. Traffic accidents associated with the storm injured forty people and caused $500,000 in property damage.

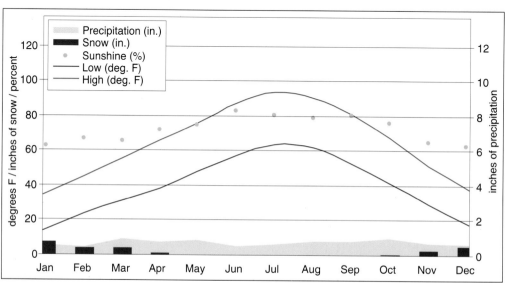

Climate of Grand Junction, Colorado

10

Colorado Plateau and Southern Rocky Mountains

Elevation: 2,500 to 13,500 feet

Natural vegetation: needleleaf evergreen trees in the higher elevations; broadleaf evergreens, sagebrush, and other dwarf shrubs at lower elevations (generally less than 5,000 feet)

Köppen climate classification: Highland in the higher elevations; semiarid or steppe at the lower elevations (generally less than 5,000 feet)

Most dominant climate feature: scattered thunderstorms in the late summer and cool winters

January brings the coldest temperatures and the most snow of the year in this region. Nights can get very cold, with readings dropping below zero in the mountains and in the teens elsewhere. Daytime temperatures vary widely, from the 40s across Arizona, the 30s in western Colorado, and the 20s in the mountains. Precipitation ranges from less than 1 inch across western Colorado to 4 inches in the Rocky Mountains of northern New Mexico. Snowfall totals for January alone reach 20 inches in Flagstaff, and the mountains receive over two times that amount. Between storms, sunny blue skies prevail.

During February and March, the storm track occasionally dips into this area, bringing rain and snow. March enters as one of the wettest months in the southern Rocky Mountains and Colorado Plateau, but by late March the storm track slowly begins to migrate north, allowing temperatures to moderate and reducing the amount of precipitation. The average precipitation for February and March ranges from ½ inch in the lowlands of southwestern New Mexico to 10 inches in the mountains of northern Arizona and northern New Mexico. As the days grow longer temperatures warm quickly, marking the end of winter. By late March, average daytime temperatures reach into the 50s. Overnight lows in February range from the teens to the 20s, but warm into the 20s and 30s in March; however, any lingering winter snow cover can cause nighttime temperatures to drop into the teens.

In April springtime weather arrives with temperatures reaching into the 50s and 60s and overnight lows generally in the 30s. Rapidly warming temperatures cause rivers to swell with snowmelt, which sometimes leads to flooding. Snow, if any, only falls at the highest elevations (generally above 8,000 feet) except during surprise winterlike storms. The average April precipitation ranges from a few tenths of an inch over southwestern New Mexico to 2 to 4 inches in the Arizona and New Mexico mountains.

As the storm track continues to move north in May and June and the strong spring sun warms the air, precipitation amounts drop dramatically and temperatures increase rapidly. By June daytime temperatures climb into the 80s and skies are usually clear. Morning lows warm from the

30s and 40s in May to the 40s and 50s in June. The cooler summertime temperatures in higher terrain attract those wishing to escape the intense heat in the lower deserts. Occasional intrusions of moisture from the Gulf of Mexico help form scattered thunderstorms, which drop welcome rain in the western portions of the southern Rocky Mountains, while drier conditions prevail across Arizona and southeastern Utah. May and June are the driest months of the year in this region, with a few tenths of an inch of rain across most areas, but up to 4 inches in the western portions of the southern Rocky Mountains.

Not until late July or August does widespread rain become possible with the onset of the Southwest monsoon (see *Southwest Monsoon* in *chapter 6*). Tropical moisture from the Gulf of Mexico and the Pacific Ocean off the west coast of Mexico converge over this region. The result, daily scattered thunderstorms, often produces spectacular lightning, flash flooding, and small hail. Most areas in this region experience their wettest months of the year during July, August, and early September. On average up to 12 inches of rain falls in the higher mountains of Arizona and New Mexico during July and August. The lowlands across southern Arizona and New Mexico receive up to 6 inches, and up to 2 inches fall across western Colorado. Despite the increased rainfall, temperatures manage to rise to their highest levels of the year in July, but overnight lows tend to be very comfortable due to the high elevation. Daytime highs reach into the 70s in the mountains and 80s and 90s at lower elevations. Daily highs in the 100s commonly visit the deserts of southeastern Utah. Low temperatures tend to be cool but comfortable in the 50s and 60s.

The summer monsoon begins to wane in September, while cooler autumn air moves into the region by October. Rainfall averages drop, and the amount of sunshine increases. September, slightly wetter than October across the entire southern Rocky Mountains, ushers in two-month average precipitation totals reaching 6 inches in the higher mountains in south-central New Mexico to 2 to 4 inches elsewhere. Temperatures can still rise oppressively in September, but they quickly cool to comfortable levels in October. Daytime highs cool from the 70s and 80s in September to the 60s in October. Clear skies, dry air, and longer nights allow nighttime temperatures to cool into the 30s in October.

Precipitation continues to diminish through November and December. The two-month average precipitation ranges from 1 to 3 inches, but up to 6 inches in higher elevations. Most of the precipitation comes as snow, but valley rain commonly falls in southern New Mexico. Temperatures drop to near their coldest level of the year and the northern branch of the jet stream moves south into Arizona, bringing with it colder air and stormy weather. Average daytime temperatures generally range in the 40s and 50s (30s in the mountains), while morning lows fall into the teens and 20s. Slow-moving Pacific storms can bring weeklong periods of heavy snow and high winds. Although not as wet as the monsoon season of late summer, winter is the second wettest season for the southern Rockies and Colorado Plateau. In between the occasional stormy periods, spells of very sunny and pleasant weather commonly occur, making this a popular destination for skiers and other outdoor enthusiasts.

Where Has All the Snow Gone?
San Francisco Peaks, Arizona

North of Flagstaff, rising more than 12,000 feet above sea level and also known as San Francisco Mountain, the "peaks" occur on a giant stratovolcano composed of many layers of lava and ash that accumulated over millions of years from thousands of eruptions. The mountain has several different peaks, which line a ridge that forms a west-facing bowl. This mountain, one of the largest in Arizona, covers 2,000 square miles and has the highest point in the state—12,600 feet.

By the time spring arrives, the snowpack in these mountains often measures 100 inches or more, enough to support a winter ski resort called Arizona Snowbowl. The resulting snowmelt produces by far the most important natural resource in the southern Rocky Mountains and Colorado Plateau region, because it provides water to the thirsty cities of southern California, southern Nevada, and Arizona. A large portion of the water goes toward irrigating the valleys of southern California, some of the nation's crucial agricultural regions. However, much of this snow disappears each winter and spring in a process called *sublimation,* before it can enter a river or the water table. One former Wyoming Soil Conservation Service Snow Survey supervisor called evapo-sublimation "a thief that steals our water."

During sublimation, snow and ice (solids) change directly to a gaseous state, skipping the liquid phase altogether. During deposition, the opposite of sublimation, water vapor turns into precipitation. Chemists use the term *deposition* when a vapor becomes a solid. American meteorologists often use *sublimation* to describe both processes. Energy released or absorbed during a change of state is called *latent heat;* for example, energy released when water turns to ice, or energy required to change water to water vapor. Heat energy needed for sublimation is referred to as *latent heat of sublimation.* The sun provides the most abundant source of this latent heat, and all locations with a snowpack and sunny skies experience sublimation. Unlike the cloud-covered Cascades in Oregon and Washington, the abundantly sunny mountains of northern Arizona lose a significant amount of snow through sublimation.

Since snow often melts and then evaporates at the same time during sublimation, we call the combined water loss from evaporation and sublimation *evapo-sublimation.* A snowpack melts and sublimates at the same time, making it difficult to measure accurately the amount of snow lost to just sublimation; however, scientists do make estimates using laboratory devices. Field measurements do not provide the controlled environment necessary to monitor these processes. Because of the complex nature of evapo-sublimation, scientists continue to work to unravel its mysteries. We do know, however, of the profound effect it has on runoff rates.

Researchers at Flagstaff's Northern Arizona University, who have studied the San Francisco Peaks, estimate that at least 22 percent to as much as 70 percent of the annual snowpack's snow-water equivalent is lost to sublimation and evaporation each year. Evapo-sublimation occurs any month that snow covers the ground. On a daily basis the San Francisco Peaks lose between 0.04 and 0.39 inches of a snowpack's snow-water equivalent; the daily mean reaches about .08 inches. These seemingly small amounts quickly add up over winter and spring and constitute a large amount of water loss. For example, the White Horse Lake weather station, high in the San Francisco Peaks, loses an estimated 8 and 14 inches of precipitation each year to evapo-sublimation. That equates roughly to 28 to 49 percent of this location's average annual precipitation of 28.82 inches.

Low humidity, moderate winds, and abundant sunshine create the optimum weather conditions for maximum evapo-sublimation. Moderate winds provide turbulence and mixing. This allows drier air, which has the capacity to absorb more moisture, to replace moistened air nearest the surface. Arizona's snow falls prey to sublimation because of the state's arid climate: average annual afternoon relative humidity at Flagstaff, which is representative of Arizona's climate, reaches 38 percent; average winter wind at Flagstaff blows nearly 8 miles per hour (much stronger at the mountaintops);

and Flagstaff ranks as the sixth sunniest U.S. city, receiving 79 percent of its total possible annual sunshine.

Residents and scientists often overlook this silent and invisible act of nature, but it has huge implications for the water supply of the southwestern United States. Less snowmelt means less runoff for rivers. This means less-than-ample water availability to support the needs of municipalities and agriculture. In 2001, during the peak of the 1999–2004 drought, the Southwest eagerly awaited the spring runoff. Given the below-average (33 percent of average) snowpack and snow-water equivalent in the southwestern mountains, farmers did not expect rivers at normal flow levels, but they knew that any runoff would help their wilting crops. However, rivers ran lower than normal that year because of evapo-sublimation. For example, between April and June, the average monthly stream flow of the Verde River below Bartlett Dam in Arizona measured 50 percent of average (308 instead 603 cubic feet per second). Some rivers, such as the Gila River below Painted Rock Dam, didn't flow at all during much of May and June. The average stream flow on the Gila this time of year is 771 cubic feet per second! It was as though the snowfall had disappeared.

New Mexico Snow: Sandia Crest

From the Colorado Plateau in the northwestern part of the state to the Great Plains of its eastern border, New Mexico is topographically, geologically, and culturally diverse. As one might expect in a place that has desert lowlands, volcanic highlands, plateaus, and high peaks, New Mexico's climate offers diversity; it has areas that receive less than 5 inches of rain a year but also towering mountains that collect over 200 inches of snow each winter. Data from SNOTEL stations at different locations lead meteorologists to believe that the Sangre de Cristo Mountains, 25 miles south of Taos, receive an average 222 inches of snow a year.

Many people probably think of New Mexico as a dry, dusty place, and they are right. The southwestern and south-central portions of New Mexico—the driest areas in the state—receive 6 to 8 inches of precipitation a year. The landlocked deserts of southern New Mexico lie far from a direct source of moisture in Earth's desert belt—the area near 30 degrees north latitude. In this region of the world, the global wind system directs air downward, forcing it to dry out and warm up. Dominated by high pressure (descending air), this area of the United States is rarely visited by wet storms. Furthermore, southern New Mexico is affected by the rain-shadow effects of the Rocky Mountains to the west and smaller mountain ranges, such as the Sacramento and San Andres Mountains, to the east.

Over half of the precipitation that does fall on southwestern and south-central New Mexico comes from late summer and early fall thunderstorms associated with the Southwest monsoon (see *chapter 6*). The eastern plains of New Mexico, however, receive relatively heavier precipitation (generally 15 to 20 inches a year) as a result of low-pressure areas that re-form after hitting the mountain ranges. These storms tap into low-level moisture from the Gulf of Mexico and rotate it counterclockwise into eastern New Mexico, resulting in heavy spring and summertime rain and thunderstorms.

The potential evaporation of southwestern and south-central New Mexico—the amount of water that theoretically could evaporate if it were available—far exceeds the total annual precipitation. This situation is termed a *water deficit*. In southern Arizona and New Mexico more than 100 inches of water could evaporate in a single year, and with an average 10 to 20 inches of rainfall a year, this means southwestern and south-central New Mexico has a 90-inch water deficit. A water deficit of this magnitude partly defines an arid or desert environment. It limits the amount of water available for vegetation, meaning only drought-tolerant species such as cactus, sagebrush, and pinyon pine can survive. For people, a water deficit means getting water from other areas to sustain irrigation and other needs.

In 1910, a weather station 30 miles southwest of Deming reported only 1 inch of precipitation, making it the driest year ever at a New Mexico location. Southwest of Deming the Animas and Hatchet Mountains, which soar to 8,532 feet, collect upwards of 30 inches of precipitation a year. The seasonal streams that trickle from these high mountains make up an important water supply for the lowlands, including Deming. Not all years lack moisture, though.

One of the most interesting characteristics of the semiarid climate of southern New Mexico is the extreme variability of its annual precipitation. While bone-dry some years, the state can get extraordinarily wet. As with most arid regions, tremendous variability in precipitation over time

is normal. Take for example Las Vegas in north-central New Mexico, which measured only 6.29 inches of precipitation in 1956 but had a phenomenal 32.29 inches in 1941. New Mexicans can't really count on "normal" yearly precipitation.

New Mexico's Wettest and Driest Spots

At the driest spot in New Mexico, near Shiprock—close to the Four Corners—only 7.50 inches of precipitation falls, on average, a year. At 8,663 feet in the Sacramento Mountains, Cloudcroft, just 25 miles east of White Sands and the wettest city in New Mexico, receives an average 25.70 inches of precipitation a year. The wettest location in New Mexico, atop the Sangre de Cristo Mountains northeast of Santa Fe, gets 45 inches of precipitation each year.

White Sands National Monument —Photo courtesy of the National Park Service

Historic Snowstorms and Blizzards in New Mexico

November 2, 1946: A heavy one-day storm across the northern half of New Mexico dropped up to 3 feet of snow in the mountains and up to 6 inches in the lowlands.

February 1–5, 1956: A record snowstorm in New Mexico and west Texas produced 15 inches of snow at Roswell, New Mexico—where it snowed for ninety-two consecutive hours—and up to 33 inches in the Texas Panhandle; both of these locations receive an average annual snowfall of only 12 inches. Heavy drifting over the eastern plains made most roads impassable. The event closed schools, stranded motorists, and made caring for livestock extremely difficult. The National Guard had to aid in the rescue of stranded motorists and airlift supplies to snowbound ranches.

March 22–25, 1957: One of the worst spring blizzards on record blew snowdrifts into 10-foot heaps in northeastern New Mexico. Union and eastern Colfax Counties received 2 to 16 inches of snow, but drifts exceeded several feet. An estimated 12,000 to 15,000 head of sheep and cattle died of hypothermia. In the southern and western parts of New Mexico, the storm whipped up severe dust storms that damaged crops.

December 12–20, 1967: Regular snowfall, cold temperatures, and strong winds thrashed western New Mexico. In places the record snowfall collected into 10- to 20-foot drifts, isolating towns for days. Snow vehicles and airlifts took food to stranded families and livestock, including those on remote Indian reservations.

November 29, 1975: Thirty-four inches of snow in twenty-four hours buried Red River, in the Sangre de Cristo Mountains, establishing the record twenty-four-hour snowfall amount for New Mexico. (A March 12, 1978, snowstorm left Red River with a one-day snowfall record of 36 inches, which ranks as the current New Mexico record.)

February 3–6, 1989: A winter storm over the southwestern United States buried the ski resorts of northern New Mexico with up to 7 feet of snow in three days.

New Mexico's Big Blizzard

A tremendously powerful and large upslope winter storm hit New Mexico between December 12 and 14, 1987. It broke snowfall and wind records while stranding over one thousand motorists along New Mexico's roads. Gascon, 32 miles northeast of Santa Fe, reported a record daily snowfall of 18 inches on December 13, with a total storm snowfall of 22.2 inches. South of there, Ruidoso received 29.3 inches of snow from the storm. Winds of 30 to 60 miles per hour with gusts up to 71 miles per hour whipped snow into a deadly blizzard at Albuquerque—so severe that one person died of exposure while attempting to walk 1 mile to his home after his truck stalled. His body was found just 300 yards from his home. Snowfall from the storm totaled up to 3 feet in parts of the Sangre de Cristo Mountains and up to 2 feet at the lower elevations.

Some of the strongest winds ever measured in New Mexico blew during this storm. Unofficially, winds howled at a constant 75 miles per hour, with a record gust of 124 miles per hour recorded at about 2:30 PM at the Sandia Peak Tramway at the base of Sandia Mountain northeast of Albuquerque.

Most people don't consider New Mexico a snowy place, but it can receive rare, major winter storms, which deliver some of this country's wildest weather. The Sandia, Sangre de Cristo, and San Juan Mountains—all part of the southern Rocky Mountains—are among the snowiest areas of the state. Though capable of dropping a lot of snow, New Mexico's snowstorms are also infrequent and short-lived.

Although winter brings the driest time of year in New Mexico, powerful Pacific storms occasionally track across the U.S.–Mexico border, generating tremendous snowstorms. As storms move inland from the Pacific Coast off California, they dump much of their moisture in the coastal and inland mountain ranges of California, Nevada, and Arizona because of orographic lifting and cooling. After a storm passes over the deserts of this region, it dumps much of its remaining moisture on the western slopes of the Continental Divide in New Mexico. Very little moisture remains in these storms for the central valleys and eastern slopes of New Mexico; however, these regions east of the Continental Divide receive the greatest amount of snow in New Mexico.

A Pacific storm grows more intense (see storm-intensification discussion in *Lake Effect Snow: The Great Salt Lake* in *chapter 9*) as it moves across the lower elevations of central and eastern New Mexico. The storm's counterclockwise circulation grows stronger and draws moisture into New Mexico from the Gulf of Mexico. When a high-pressure system is present over the northern Great Plains, its clockwise rotation and frigid air mix with the Pacific storm's counterclockwise rotation and moist air, delivering heavy upslope

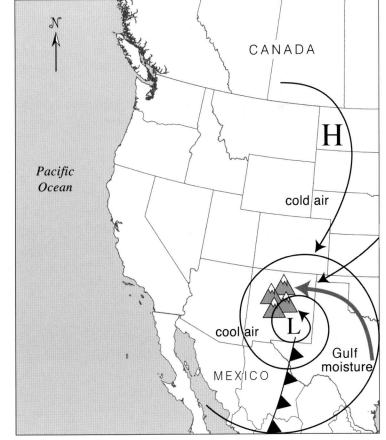

The most favorable weather conditions for heavy snow in New Mexico consist of a strengthening low-pressure area over eastern New Mexico, which draws moist air into the region from the Gulf of Mexico, and a cold high-pressure area over the northern Great Plains that rotates frigid air into New Mexico from Canada. The cold air and moisture converge over New Mexico, resulting in heavy upslope snowstorms.

snow to New Mexico. The higher elevations, particularly east-facing slopes, receive the heaviest snow since the winds blow easterly and force moisture-laden air up the mountains.

These upslope storms can deliver phenomenal amounts of snow, which accumulates to great depth at higher elevations, to a very dry state. For example, Sandia Crest (10,780 feet), 15 miles northeast of Albuquerque, measured a snow depth of 95 inches on April 4, 1973—a state record for New Mexico.

In a Flash! The Canyonlands

Nowhere else in the country does the sculpting effect of weather appear more obvious than on the Colorado Plateau. The plateau, which stretches across most of northern Arizona and into southern Utah, southwestern Colorado, and northwestern New Mexico, is a large, intact section of the earth's crust. Its sedimentary layers remain horizontal here, just as they were deposited over millions of years of geologic time. Nature's wind, rain, and freeze-thaw cycle have carved the immense sandstone domes, spires, canyons, and arches that make this area breathtaking.

Sudden and deadly flash floods have made the plateau in southern Utah and northern Arizona notorious. Arizona leads the country in flash flood–related deaths and injuries each year. Those awestruck by the canyonlands' maze of stark, narrow, sandstone beauty often forget that flash floods carved these canyons over millions of years. A heavy thunderstorm can turn a dry canyon streambed into a raging torrent of thick, silt-filled water in a matter of minutes.

By definition, a flash flood plays out in six hours or less, while more leisurely river flooding can take days or even weeks to unfold. Usually flash floods occur when extremely heavy rainfall over several hours overwhelms an area. In flat terrain, such as that of some of the deserts of the Southwest, rainwater tends to spread out and produce less severe flash floods. On the Colorado Plateau, however, mountains and canyons funnel rainwater, and the steep, nonporous topography of canyons makes them ideal for flash flooding since little of the rainwater seeps into the ground.

With terrain so varied over the West, the amount of rain necessary to induce a flash flood differs from place to place, but across northern Arizona, southern Utah, and southern Nevada, it generally takes 1 inch of rain in one hour to produce a flash flood. Desert thunderstorms, which often have juicy tropical characteristics, can unleash massive amounts of rain in a short amount of time.

The Southwest monsoon supplies the moisture for these thunderstorms (see *chapter 6*). This period arrives during late summer and early fall when winds carry moisture north, from tropical regions of Mexico, into the Colorado Plateau. Intense summertime solar radiation in this region heats the plateau's surface, which in turn conducts a great deal of heat into the air. This warm air rises, carrying tropical moisture with it, into the relatively cooler air above it. As it cools, its water vapor condenses into cloud droplets. Before long a thunderstorm develops, with its strong updrafts and downdrafts.

The mix of moisture-rich air and warm desert atmosphere creates thunderstorms similar to those of tropical regions: thunderstorms with a lot of rain but usually little, if any, hail. For a storm to become an efficient heavy rain producer, it needs a constant low-level flow of moisture, rising air, and ideally, a barrier such as a mountain to slow or stall the thunderstorms. Although a single thunderstorm will often decay after an hour or two, new storms will develop, thriving on the same ingredients that started the first storm. Prolonged periods of thunderstorm development produce extremely heavy rainfall amounts, which often produce flash floods.

The intense summer thunderstorms that occur over the Colorado Plateau are sometimes referred to as *cloudbursts,* which are sudden and heavy (equal to or greater than 3.94 inches per hour) rainfall events. A cloudburst occurred July 7, 1933, in Bryce Canyon National Park, in the High Plateaus (8,000 feet) of southern Utah. It lasted only ten minutes, but yielded 0.90 inch of rain (about 5.4 inches an hour). This storm established Utah's ten-minute rainfall record.

The Labor Day Storm of 1970

One of the most extreme rainfall events to hit the Colorado Plateau was associated, as they often are, with the remnants of a tropical storm. On September 5, 1970, extremely moist and slow-moving thunderstorms associated with Tropical Storm Norma dropped phenomenal amounts of rain on southeastern Utah, southwestern Colorado, and Arizona, with astonishing results. An eight-hour-long thunderstorm that began at about 10 AM and did not end until later that evening poured 11.40 inches of rain on Workman Creek (about 60 miles east-northeast of Phoenix at an elevation of 7,000 feet) in central Arizona, setting the state twenty-four-hour rainfall record. This record was later broken in September 1997 by an 11.97-inch deluge atop Harquahala Mountain. At 5,691 feet, this mountain reigns as the highest peak in southwestern Arizona.

Similarly, at Bug Point, Utah (24 miles east of Blanding), 6 inches fell, which long stood as Utah's twenty-four-hour rainfall record. Called the "Labor Day Storm of 1970" by Arizonans, the National Weather Service ranks this storm as Arizona's second top weather event of the twentieth century. It has the infamous distinction as the most deadly weather event in Arizona's history: twenty-three people drowned, including fourteen from flash flooding on Tonto Creek in the vicinity of Kohl's Ranch, northeast of Payson, Arizona.

The danger of flash floods comes not only from water, but also from their suddenness and lack of warning. A billowing thundershower miles away may look harmless, but it will likely lead to a flash flood somewhere in canyon country. Moreover, floodwaters from tributaries and runoff can consolidate and unleash a flash flood over two hours after a thunderstorm has ended. Narrow slot canyons that corkscrew their way through sandstone at widths of only a few feet in places are most prone to flash floods; trickles of water can quickly grow into a tremendous, confined force. These canyons can fill up with tens of feet of water quickly and lodge logs high above a once-dry creek bed. The logs and other debris remain as constant reminders of the power of flash floods. Amazingly, a place that looks like a cathedral can turn into a death trap in seconds.

Imagine the horror of trying to outrun a flash flood in a canyon as the roaring water gains on you. Sadly this happens every year in canyon country. On August 12, 1997, it took only an instant for Antelope Canyon near Page, Arizona—100 to 150 feet deep and narrow enough in places for a child to touch both its sides—to turn into a raging cauldron of chocolate brown water. In an instant eleven hikers lost their lives, stripped of their clothing by the water and buried in silt at the bottom of Lake Powell. A snapshot in a camera belonging to one of the hikers later revealed a 50- to 80-foot wall of water roaring through the canyon.

Paria Canyon, located about 15 miles west-northwest of Page, Arizona, and one of the most spectacular canyons in Arizona, ranks among the canyons most prone to flash flooding. All canyons require a different amount of runoff and rainfall to produce a flash flood. For example, a flash flood in Paria Canyon is defined as a rise in the daily runoff of at least 50 cubic feet per second, while the normal daily summer flow rate is about 5 cubic feet per second. During an average year Paria Canyon experiences eight flash floods, with

Sheep Creek Flood

On June 10, 1965, a family of seven drowned in a flash flood in Sheep Creek Canyon in Utah's Uinta Mountains. They were camped at Palisades Campground along the swollen Sheep Creek. Heavy rains turned the creek into a raging torrent, completely flooding 5 miles of the newly paved Highway 44, three recreational areas, and seven bridges. Damage estimates reached $1 million ($5.7 million in 2002 dollars). The National Weather Service ranked this flash flood as Utah's fifth most significant weather event of the twentieth century.

Flash flood debris jammed in "Buckskin Dive" of Buckskin Gulch, a tributary of lower Paria Canyon in southern Utah. Floodwaters rip through slot canyons and jam debris between canyon walls, leaving lasting reminders of a flash flood's violence. —Michael Jewell photo

Delicate Arch in Arches National Park, Utah, with the snow-covered La Sal Mountains in the distance —Photo courtesy of the U.S. Geological Survey

Generalized Flash Flood Frequency, Paria Canyon, Arizona

MONTH	AVERAGE MONTHLY PRECIPITATION	FLASH FLOOD FREQUENCY
January	0.38	1 every 5 years
February	0.47	1 every 10 years
March	0.49	3 every 10 years
April	0.39	extremely rare
May	0.31	1 every 10 years
June	0.24	rare
July	0.73	2.9 each year
August	1.18	3 each year
September	0.51	1.4 each year
October	0.42	3 every 10 years
November	0.39	1 every 2 years
December	0.44	3 every 10 years

A flash flood in Paria Canyon is defined by a rise in the daily runoff of 50 cubic feet per second or more.

Driver Beware!

Flash floods kill hikers, and they also kill motorists. Most flash flood victims (60 percent) are killed in their automobiles while attempting to drive through swift currents. Many motorists underestimate the staggering power of moving water. A foot of water running across a roadway has 500 pounds of force behind it. Two feet of water will carry away most cars, making your vehicle a threat to your life and not a source of safety. Never try to drive through floodwaters, even if you think it's shallow enough to get through, because floodwaters could have easily dredged out sections you can't see. If caught in floodwaters while still in your car, get out immediately; better wet than dead.

the highest frequency in August. The most extreme floods here produce flow rates greater than 400 cubic feet per second, which is remarkably high considering Paria Canyon's drainage basin is less than 1,500 square miles. Similar to other moderately sized canyons, the Paria's rise in runoff will normally occur over a very short time and last for one to four hours.

Second to lightning, floods—both river and flash floods—kill more people in the United States than any other weather phenomena. During a typical year nearly 140 people perish in flash floods, and nearly 50 more die in river flooding. Whenever traveling, hiking, or camping in canyon country, note any heavy rainfall or thunderstorms. Be aware that the regional National Parks do not permit hiking or camping in flood-prone canyons during periods of flash flood weather, especially during the Southwest monsoon of late summer. Never camp near a river and always keep your eyes on the sky. If threatening weather approaches, head for higher ground and closely monitor water levels near you.

The Albuquerque Box Effect

For years, before Land of Enchantment became their official state slogan, New Mexicans proclaimed that they lived in the Sunshine State. In New Mexico, a single bona fide cloudy day rarely occurs. Residents are accustomed to seeing the sun more than 70 percent of the potential daylight hours year-round. Officially known as the Sunshine State, Florida receives less of its potential sunlight than New Mexico.

While the climate varies considerably from the mountains to the plains in New Mexico, one thing is true: it is sunny. Annually, Roswell receives 74 percent of its total possible sunshine. Although remarkably sunny throughout the entire state, the Rio Grande valley is one of the sunniest and driest places in the state; the sun shines nearly 80 percent of the time there. Albuquerque sees 76 percent of its total possible sunshine. During the sunniest time of year—June and July—these cities bask in over 85 percent of their possible sunshine. This compares to Tampa, Florida, which receives 60 percent of its total possible sunshine during summer and 66 percent annually. In New Mexico, clear to partly cloudy skies prevail two hundred to three hundred days

Cloudy New Mexico Skies

Relatively cloudier skies do occur in New Mexico during El Niño winters (see *El Niño Storms* in *chapter 4*). As a result of El Niño, the strong subtropical jet stream racing across the warmer-than-normal central Pacific Ocean gathers tremendous amounts of evaporating moisture. El Niño forces the storm track to shift south, and the waterlogged jet stream travels across the southwestern part of the United States (instead of the Pacific Northwest), delivering gray skies and heavy precipitation to New Mexico. During the strong 1997–98 El Niño winter, Albuquerque endured four consecutive cloudy days, the longest such streak since December 1982, which was also an El Niño winter. Cloudier skies sometimes occur during summer and early fall when the Southwest monsoon (see *chapter 6*) provides abundant moisture for cloud and thunderstorm formation.

a year. Very few livable places can compete with such bright statistics.

This predictable sunshine partly explains why balloon enthusiasts from around the world gather yearly in Albuquerque. From a small gathering of thirteen balloons in 1972, the Kodak Albuquerque International Balloon Fiesta has grown to the largest balloon event in the world. The weather in Albuquerque, widely considered to be the best in the world for ballooning, earns the city its unofficial title as Ballooning Capital of the World. Held each year during the first week in October, the Balloon Fiesta attracts more than 850 balloons and 1,200 pilots from around the world.

You can count on the weather in Albuquerque. Crisp, calm mornings, cool brisk evenings, and sunny skies combine for ideal balloon flying conditions. The Balloon Fiesta takes place in early October for good reason. The fall weather brings light and variable winds, a stable atmosphere, and daytime temperatures in the upper 70s and morning lows near 50°F. These conditions combined with the local topography cause a unique valley wind system known as the Albuquerque Box Effect.

The famous Albuquerque Box Effect creates winds that flow in opposite directions at different altitudes and enables balloon pilots to fly with controlled precision. It even allows them to launch and

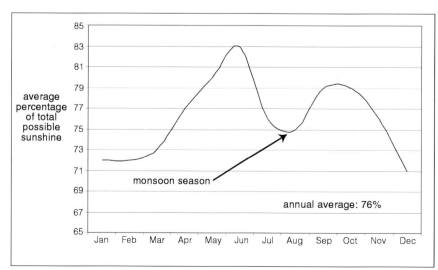

Average monthly percentage of total sunshine at Albuquerque, New Mexico, based on data recorded between 1940 and 1993

Hot air balloons at the Albuquerque, New Mexico, Balloon Fiesta —Fred Voetsch photo

land at the same location. The mountain formations and wind characteristics of the Rio Grande Valley cause this unique weather pattern.

The Box Effect develops in the Rio Grande Valley under stable atmospheric conditions. During nighttime hours, the air near the ground cools through the process of radiational cooling. This process works most efficiently with clear skies, low humidity, and light wind—common weather for the valley during early October. The mountaintops to the north and east of Albuquerque—the Sangre de Cristo, Santa Fe, Jemez, and Sandia Mountains—cool faster and to a colder temperature than the valley elevations, so cool dense air flows down off the mountains and into the valley. The cool, shallow layer of air generally measures no more than a few hundred feet deep.

During the early morning hours this air flows southward down the valley from higher to lower elevation, much as any fluid flows downhill. This creates a north wind, generally less than 10 miles per hour, in the middle Rio Grande Valley near Albuquerque. An inversion develops over the valley when warm winds aloft (about 5,000 feet above the surface) associated with the Great Basin high (see *chapter 7*) blow lightly from the south or

southeast, over the cool air near the valley floor. These opposing winds are what the balloonists are after. The winds in an ideal box pattern blow in opposite directions: a northerly wind at the surface and a southerly wind above the surface. A skillful balloon pilot can bring a balloon back to the point of takeoff by changing altitudes and riding the different wind currents.

When pilots take off at the Balloon Fiesta grounds north of Albuquerque, they first head south toward downtown Albuquerque, then ascend into the winds that take them back to the fiesta grounds. As the sun heats the ground in the morning, thermal turbulence mixes the separate layers of air, destroying the low-level inversion by midmorning and ending the Box Effect.

Strong winds sometimes cancel balloon flights, but rarely in New Mexico during early October. The average wind in Albuquerque during October blows 8.2 miles per hour. Winds below 10 miles per hour are ideal; winds of 10 to 15 miles per hour make launches very difficult; and launching in winds above 15 miles per hour is virtually impossible. Without New Mexico's mild climate, this spectacular event, which draws worldwide attention, would not be possible.

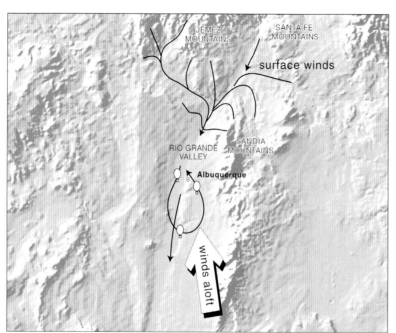

Ideal weather conditions for the Albuquerque Box Effect. Cool surface winds drain out of the higher mountains and cause northerly winds in the Rio Grande Valley. Winds aloft (5,000 feet above the ground) blow from the south over the cool air, creating an inversion and the wind pattern balloonists love. A skilled balloonist can bring a balloon back to the point of takeoff by changing altitudes and riding the different wind currents

11

Southern Great Plains and Adjacent Foothills

Elevation: 2,500 to 8,000 feet

Natural vegetation: needleleaf evergreen trees at elevations generally above 6,000 feet; grass and other herbaceous plants at lower elevations

Köppen climate classification: dry, particularly in the summer, with cool temperatures

Most dominant climate feature: frequent afternoon and evening summertime thunderstorms

The southern Great Plains and adjacent foothills enjoy mild winters. The coldest temperatures for this region occur in January and February when daytime highs reach into the 50s in the south and 40s in the north. Morning lows typically drop into the 20s. A Chinook (see *Snow Eater* in *chapter 12*) flowing off the Rockies mixed with sunny skies, however, can produce a string of 70°F days in the dead of winter. Precipitation, which falls mostly as snow—except in the extreme southern reaches—totals only 1 inch or less during January and February.

By March and April strong Pacific storms tracking across southern Colorado can deliver heavy snowstorms, especially along and near the foothills of the Rocky Mountains. Total precipitation accumulation ranges from about ½ inch over the eastern plains to 3 inches in the foothills. Temperatures continue to climb, and by April daytime highs reach the 60s in the north and

70s in the south, with overnight lows mostly in the 20s and 30s in March and the 40s in April. The last spring freeze for the southern part of this region often arrives mid-April.

The northern stretch and foothills of this region receive their last spring freeze in early to mid-May. May and June are among the wettest months of the year here, with frequent cold-front passages and thunderstorms delivering 1 to 4 inches of precipitation across the region. This time of year also marks the beginning of hail season; frequent hailstorms often occur with the almost daily thunderstorms. Daytime temperatures warm into the 80s and 90s by June with overnight lows dropping to the 50s in the north and 60s south. The first 100°F heat often occurs in June.

Temperatures continue to warm in early July then level off in August. Daytime highs typically peak in the 90s, with lows only in the 60s. Strings of days with 100°F heat commonly occur, but afternoon thunderstorms often provide a cool rain, especially near the foothills. Some of the thunderstorms, however, reach severe limits—with winds greater than 57 miles per hour or hail ¾ inch in diameter or larger—at this time of year, causing tornadoes, damaging winds, large hail, and heavy rainfall. Total precipitation for July and August combined can exceed 5 inches in the south, but most areas receive between 1 to 4 inches.

By mid-September in the north, and late October in the south, the first fall freeze occurs;

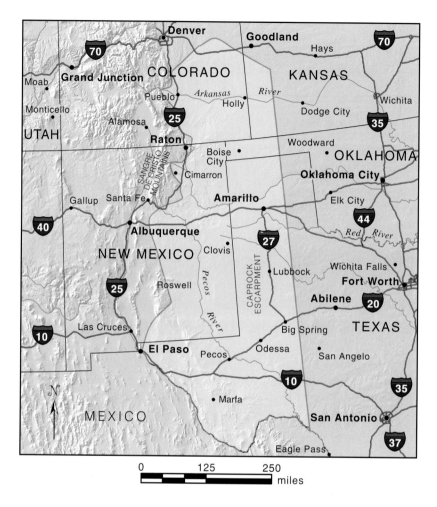

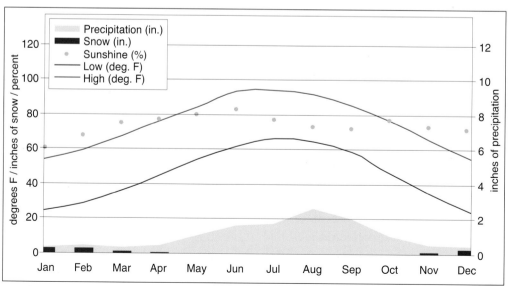

Climate of Roswell, New Mexico

otherwise, average nighttime lows warm to the 50s during September, cooling into the 40s by October. Daytime highs remain mild, with readings in the 65 to 75°F range.

Although the foothills can see measurable snow by early November, most areas get their first—if any—snowfall in December. Snow in the extreme southern reach of this region rarely falls any time of the year. Most snow falls in the foothills when cold air from the north mixes with moist air from the southeast as it blows up against the foothills. These upslope storms can leave the foothills of northern New Mexico and southern Colorado under several feet of snow, while the lowest elevations receive cold rain. Although big storms can happen, average total precipitation in this region only ranges from a trace to locally 2 inches. Daytime temperatures hover mostly in the 50s with morning lows dipping into the 20s in the north and 30s in the south.

Thunderstorm Alley of the West

Second only to Florida, the Rocky Mountain foothills adjacent to the southern Great Plains receive more thunderstorms annually than anywhere else in the United States. Residents in north-central New Mexico and southeastern Colorado hear the rumbles of thunder 60 to locally 110 days a year. Known by locals as the "thunderstorm capital of the West," Cimarron, New Mexico, east of the Sangre de Cristo Mountains, endures more thunderstorms than anywhere else west of the Mississippi River. It averages 110 thunderstorms a year—one thunderstorm every day from June 1 to September 8.

Several climatic and topographical features make the foothills of northern New Mexico an excellent breeding ground for thunderstorms. Along the foothills, cool, dry upper-level westerly air typically keeps skies clear and allows the sun to heat the surface rapidly, which creates an area of warm rising air. (Eventually, this same upper-level air pushes thunderstorms east.) Low-level

Locations with Greatest Frequency of Thunderstorms

RECORD	LOCATION	THUNDERSTORM DAYS PER YEAR
Greatest in world	Bogur, Java (Indonesia)	322
Greatest in U.S.	South-central FL	120
Second greatest in U.S./ Greatest in western U.S.	Cimarron, NM	110

moist air flows over the southern Great Plains from the Gulf of Mexico. Higher pressure over New Mexico and relatively lower pressure over southern Texas causes this low-level moisture, sometimes referred to as a *low-level jet stream.* The two opposite circulations around these systems funnel the moisture into the area. Often the atmosphere is stable (warm air aloft) across Texas, and it won't allow the moist air mass to develop clouds; however, when this air mass hits the foothills and the Sangre de Cristo Mountains with their rising superheated air, clouds do form.

With their southwest to north-northeast orientation, the Sangre de Cristo Mountains in north-central New Mexico soar to elevations of 12,440 feet and provide the moist air mass an optimum ramp to ride. As the mountains and the superheated air force the moist air to rise, its water vapor cools and condenses on dust particles, turning into water droplets. Since plenty of dust particles live in this arid environment, millions of microscopic water droplets form and create a cumulus cloud.

When water vapor transforms into water droplets or cloud particles, it releases large amounts of latent heat. This keeps the air inside the cloud warmer than the air surrounding it, forcing it to continue expanding horizontally and vertically. Warm, moist air continues rising, feeding the cumulus cloud. This is called the *updraft,* and its formation is the beginning of a thunderstorm. Cloud droplets combine to form raindrops (or ice crystals at the higher cloud levels) within the cumulus cloud. As the cloud builds well

Types of Thunderstorms

Single cell: A storm with one updraft and one downdraft, generally lasting twenty to thirty minutes and usually not strong enough to produce severe weather; able to produce brief heavy rain and gusty downdraft winds.

Multicell: A cluster of storms with several updrafts and several downdrafts in close proximity; the most common type of thunderstorm, multicells can produce brief heavy rain, moderately sized hail, strong winds, and weak tornadoes. Although each cell in a multicell cluster lasts only about twenty minutes, the multicell cluster itself may persist for several hours. When these clusters grow to sizes exceeding 180 miles in diameter, meteorologists call them *mesoscale convective systems* or MCSs.

Multicell line: A long line of storms with several updrafts and several downdrafts; often called a *squall line* by meteorologists. Squall lines can produce severe weather, particularly large hail, brief heavy rain, weak tornadoes, and strong downburst winds.

Supercell: A highly organized thunderstorm with strong updrafts in a multicell cluster or line; although rare, supercells can produce very large hail, heavy rain, and most importantly, violent tornadoes.

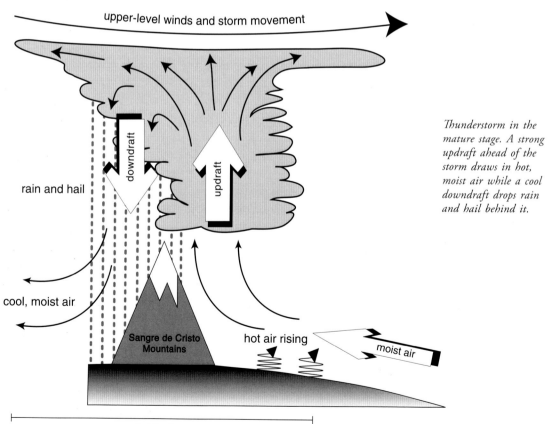

Thunderstorm in the mature stage. A strong updraft ahead of the storm draws in hot, moist air while a cool downdraft drops rain and hail behind it.

above the freezing level of the atmosphere the raindrops can no longer remain suspended, and they fall. Some raindrops evaporate as they fall, cooling the air beneath the cloud. This pocket of dense, cold air builds and rushes to the ground as a downdraft. The downdraft's speed increases as falling precipitation drags the air down. When the updraft and downdraft occur within the same cloud, rather than within a series of clouds spread over a certain distance, the storm is termed a *cell* or *thunderstorm cell*. At this stage, with both

Remarkable Tropical Remnants

Every twenty years or so, massive amounts of moisture blow into New Mexico from decaying Gulf of Mexico tropical storms and hurricanes, causing large areas of flood-producing rain and thunderstorms.

On August 31, 1942, the remnants of an unnamed hurricane tracked across the Gulf of Mexico, Texas, and into New Mexico. Remarkable in several ways, the dying hurricane quickly reached the lower Pecos Valley of New Mexico within thirty hours of its landfall in the vicinity of Matagorda Bay in Texas. The storm still had a well-defined cyclonic circulation when it hit New Mexico. It resulted in record-breaking twenty-four-hour rainfall amounts at most weather stations in the watersheds of the upper Pecos and Canadian Rivers. Raton received 5.75 inches in twenty-four hours. Not a single station in the path of the storm reported thunder. This meant the massive slug of moisture and the instability associated with the tropical low caused the rain and not moisture supplied by a low-level jet stream from the Gulf of Mexico or orographic lifting.

Other tropical storms have moved into New Mexico: an unnamed hurricane, August 21, 1886; and Hurricane Gilbert, September 15, 1988.

an updraft and a downdraft, the thunderstorm peaks. Lightning can develop at this point as the electrical field within the storm and the surface begin to differ (see *Lightning Capital of the West* in *chapter 6*).

Besides mountains, dust, and warm, moist air, this region's ample sunshine plays a key role in the production of thunderstorms. Without the hindrance of clouds or haze, sunshine quickly heats the surface of the foothills, which in turn heats the air above, creating a column of hot, rising air in a chimney of relatively cooler air. This promotes efficient convection—the rising of air—which thunderstorms require. Storm clouds eventually block sunshine, but not before storms have grown strong enough to no longer require the warm air near the surface. Instead, a storm's updraft draws in the moisture and heat it requires. As storms die, new ones are produced along gust fronts—cool rushes of air caused as downdrafts push outward from decaying storms.

Like clockwork, daily thunderstorms burst out of clear air at about 1:00 or 2:00 PM, then dissipate or move eastward into Texas by sunset. Even without a lot of atmospheric support—moisture and instability—scattered thunderstorms dot the afternoon sky. These predictable daily thunderstorms are referred to as *diurnal thunderstorms*.

Typically, multicell thunderstorm events in the southern Rocky Mountain foothills will only last 70 to 80 minutes, but some of the better-organized storms—those constantly resupplied by moisture—move eastward and promote the development of more thunderstorm cells, leading to thunderstorm events lasting 100 to 110 minutes. An individual thunderstorm event lasting more than 100 minutes is quite long, and only two areas of the country usually experience such long events: the central Great Plains and the Colorado Plateau. These areas experience relatively long-lasting thunderstorms events for different reasons. During the Southwest monsoon (see *chapter 6*), Colorado Plateau thunderstorms

A beautiful supercell thunderstorm over Clines Corner, New Mexico, near Interstate 40 —Charles Bustamante photo

receive an abundant supply of moisture from the south, which allows these storms to last a long time. Great Plains thunderstorms often cluster together and form long-lasting mesoscale convective systems, or MCSs, which last several hours and can be as large as 300 miles wide.

Too Many Thunderstorms to Count

At any one moment nearly two thousand thunderstorms rumble over the surface of the earth, mostly in and near the tropics. During a single year the United States sees an estimated one hundred thousand thunderstorms, and about one thousand of these spawn tornadoes.

Since strong upper-level winds often move New Mexico thunderstorms eastward quickly, they do not have time to drop copious amounts of rain on any one spot, unlike places in the Colorado Plateau. Often the storms only drop virga—rain that evaporates before reaching the ground—since the air is so dry. Even with eighty thunderstorms a year, the Rocky Mountain foothills in New Mexico only receive 15 to locally 25 inches of precipitation a year—about 55 percent of that from thunderstorms. Florida, with 80 to 110 thunderstorms a year, receives 45 to 65 inches of rain annually—about 60 percent of the region's annual rainfall comes from thunderstorms. Seventy percent of the Texas Panhandle's annual precipitation comes from thunderstorms, which makes it the highest percentage in the United States; the West Coast receives only 10 percent of its annual precipitation from thunderstorms.

Occasionally afternoon thunderstorms become severe and produce damaging hail, gusty winds, frequent lightning, and heavy rain. Although rare, thunderstorms can spawn tornadoes across the northeastern plains of New Mexico, but not in the foothills area. A particularly severe series of thunderstorms occurred between June 14 and 17, 1965, bringing hail and heavy flooding rains to northeastern New Mexico. Raton, near the Colorado border, reported 10.35 inches of rain. Property damage reached the tens of millions of dollars. However, most thunderstorms in the New Mexico foothills, generally weak, come and go with a welcomed shower, puff of wind, and crack of thunder.

Colorado Low

The high-reaching mountain ranges of the West have a profound influence on large-scale storms. Pacific storms that slam the West Coast must cross the Sierra Nevada and Cascades, the Great Basin, and then the 1- to 2-mile-high, 500-mile-wide Rocky Mountains before reaching the Great Plains. As storms cross this barrier, the mountains force them into a smaller layer of the atmosphere.

Bound by the mountains underneath and the tropopause above—the top of the troposphere, which storms can't penetrate—the storms spread horizontally, weakening their counterclockwise circulation. The rough topography of the Rocky Mountains, Cascades, and Sierra Nevada literally rip storms apart, particularly the lower levels of a storm. Only the strongest storms can withstand these rocky forces and maintain some of their storm characteristics.

Once a storm, or the remnants of one, reaches the eastern side of the Rocky Mountains, it can reconstitute itself in the smoother topography of the Great Plains. Here a storm has enough vertical space to spin itself into a compact, 200- to 600-mile-diameter cyclone. Like an ice skater pulling her arms toward her body, a storm spins faster as it grows vertically. Pacific storms often gather themselves together in southeastern Colorado, and meteorologists refer to these storms as *Colorado lows,* or *Colorado lee cyclones.*

A low-pressure area called a "Colorado low" grows strong over southeastern Colorado after crossing the Continental Divide. Its counterclockwise circulation draws moisture into Colorado from the Gulf of Mexico. At the same time, clockwise winds blow artic air into the region from a high-pressure system over the northern Great Plains. The moisture-laden air and the artic air clash over Colorado, producing heavy snow along the Front Range and a large area of rain and snow—known as the precipitation shield—*over portions of Colorado and surrounding states. Colorado lows typically take a day or two to travel across Colorado, and then they move slowly northeast, spreading heavy rain and snow across the central Great Plains. The sweeping lines across the map are streamlines, which show the upper-level wind pattern.*

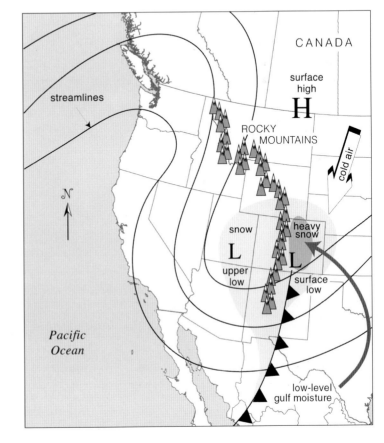

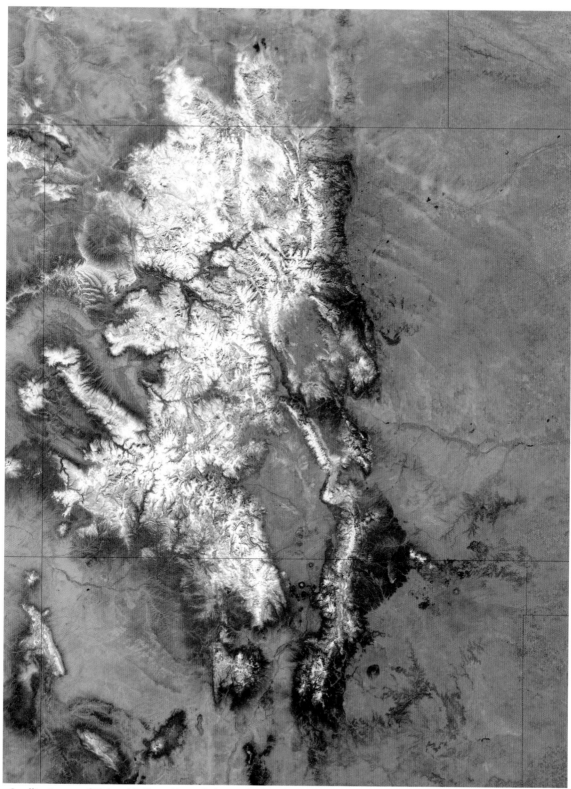

Satellite images of Colorado's Front Range before and after the "Blizzard of 2003." The image above was taken on March 13. The image on the next page was taken on March 20 after Denver received 30 inches of snow and other Front Range locations received over 7 feet. —Images courtesy of Jacques Descloitres, MODIS Rapid Response Team, NASA/GSFC

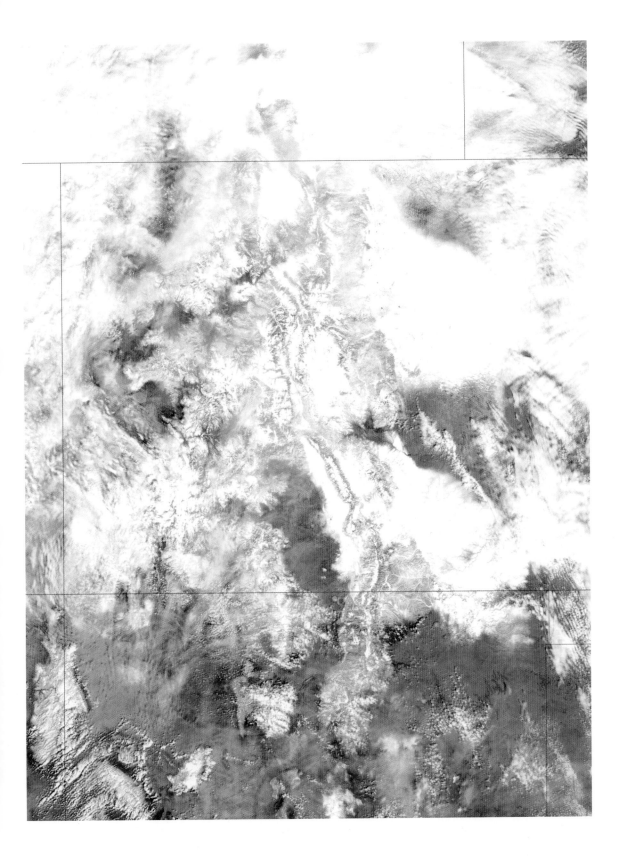

A car buried by 30 inches of snow along a residential street in Fort Collins, Colorado. The March 18–20, 2003, blizzard dropped 20 to 80 inches of snow along the Front Range.

With a Colorado low centered over south-eastern Colorado, the storm's trailing cold front can tap tremendous amounts of moisture from the atmosphere above the Gulf of Mexico. The cold front travels with a Pacific storm throughout its journey across the West, forming a line that delineates where a cold air mass and a warm air mass meet. The cold front has intense precipitation, with warm air ahead of it and colder and drier air behind it. As the moisture associated with the low's circulation mixes with arctic air plunging southward from a high-pressure area over the northern Great Plains, it cools and condenses. The counterclockwise circulation of the low slings this low-level moisture over the Front Range, forcing the air higher, where it cools and condenses even more. Eventually the air mass can no longer hold the moisture, and it dumps its load, often as snow.

The higher the water content of the moisture stream from the Gulf of Mexico, the longer a storm persists, and the heavier the snowfall. During El Niño winters, the storm track shifts south—a favorable position for the development of Colorado lows. Plus, storms tend to contain more water vapor since they have not only gathered moisture from the Gulf of Mexico but also from El Niño–warmed waters over the central Pacific Ocean (see *El Niño Storms* in *chapter 4*). Some of the most significant Front Range blizzards have occurred during El Niño winters.

Colorado lows are often associated with up-slope storms in the foothills. Occasionally they form quickly enough to spread heavy precipitation across the lower elevations of the Front Range of Colorado before sweeping violently across the Midwest, unleashing a potpourri of severe and wintry weather. Once a mature Colorado low moves into the Texas Panhandle, it sometimes gets the nickname "Texas Panhandler." Colorado residents know from experience that an east to northeasterly wind anytime of the year, particularly in the spring, will almost always bring precipitation to the Front Range.

In the spring, what meteorologists call a *Four Corners low* can precede a major Colorado low. Named after the area in which it intensifies—the intersection of Utah, Colorado, New Mexico, and Arizona—a Four Corners low almost always signals a significant snowstorm for eastern Colorado and New Mexico. A Four Corners low delivers heavy mountain snow and low-elevation rain to Arizona and Utah. As a Four Corners low intensifies, it generates a great deal of contrasting surface pressure across the Southwest, also delivering strong wind to Arizona and Utah.

After a Pacific storm has overtopped the Sierra Nevada, it grows stronger as it crosses southern Nevada. The relatively flat topography across northeastern Arizona, southern Utah, and northern New Mexico—a reconstituted storm's path—creates little resistance for a storm and

Significant Blizzard-Producing Colorado Lows in the Denver Region

December 4–5, 1913: The Great Storm dumped nearly 26 inches of snow on Denver and another 20 inches the next day. Snow totals for the two-day storm measured 45.8 inches. The climatological records reported that "livestock and transportation interests suffered greatly."

September 20, 1965: A deadly early-season snowstorm killed twelve people across Colorado, Nebraska, Wyoming, and Montana. It stranded 6,500 motorists along U.S. 30 in Wyoming in what the National Weather Bureau called the "worst early season snowstorm in the history of the West." Temperatures in the 20s and 30s, combined with heavy snow (generally 1 to 2 feet) and high winds caused blizzard conditions.

December 23–24, 1982: The worst winter storm in seventy years hit Denver. The infamous "Blizzard of '82" brought the area to a standstill, dumping 1 to 3 feet of snow. A very localized storm, it left only a dusting in Fort Collins and Cheyenne, Wyoming, while burying Denver and Boulder.

April 2–3, 1986: A major spring blizzard delivered 1 to 2 feet of heavy, wet snow and winds up to 70 miles per hour to the Front Range. Sections of all interstates leaving Denver closed at times. The snow

came after an unusually warm late winter and early spring. The trees already had leaves so the heavy snow accumulated in them and knocked them down. Trees fell on power lines, creating widespread and prolonged power outages. The storm claimed three lives, all due to exposure.

October 24–25, 1997: The worst October winter storm since 1923 occurred. The ferocious twenty-four-hour, El Niño–powered blizzard dumped 4 feet of snow in northeastern Colorado and along the Front Range. Winds blew up to 50 miles per hour and produced 15-foot snowdrifts and whiteout conditions. Ranked as the second heaviest snowstorm in Denver's history, it dropped 21.9 inches of snow.

March 18–20, 2003: The "Blizzard of 2003," the heaviest snowstorm to hit the Colorado Front Range in ninety years, dropped 20 to 80 inches of snow in the Denver–Fort Collins corridor. The weight of the snow, which had a snow-water equivalent of 5 inches of water per 3 feet of snow, caused roofs to cave in at many businesses. The storm's precipitation significantly affected the water supply amid a severe three-year drought. Both Rollinsville and Fritz Peak, southwest of Boulder and in Gilpin County, received the most snow: 7.3 feet!

allows it to stay organized. Following the path of least resistance, a Four Corners low almost always moves eastward along the southern Colorado border, one of three primary storm tracks in the western United States. As the storm tracks across southern Colorado, it may weaken, but it will almost always morph into a strong Colorado low when it moves into southeastern Colorado because it can develop fully on the open plains.

The famous "Blizzard of '82," a classic example of a Colorado low, ranks as one of the worst winter storms to ever hit the Denver area. The blizzard paralyzed Denver with 1 to 3 feet of wind-driven snow between December 23 and 24, stranding thousands of travelers and killing five motorists. Perfect conditions combined to create this monumental storm. An El Niño weather pattern drove a wetter-than-normal jet stream across the southern United States. This jet stream helped create a potent Four Corners low, which stalled over northern New Mexico and eventually morphed into a Colorado low.

The Colorado low tapped low-level moisture from the Gulf of Mexico and wrapped it around the northern portion of the storm and into the cold air. This produced strong and persistent upslope flow focused on eastern Colorado and the adjacent Great Plains, where snow commonly fell at 1 to 3 inches per hour. An increasing pressure gradient between an arctic high over the northern Great Plains and the deepening low over the southern Great Plains led to strong winds and significant blowing and drifting snow.

Another classic example occurred on February 22, 1977, when a violent Colorado low kicked up huge amounts of dust that settled on the central East Coast two days later. And again, this occurred during an El Niño year. Known for their heavy precipitation, Colorado lows can produce a great deal of dust before any appreciable precipitation falls during drought years such as 1977.

The Alberta Clipper, a cousin of the Colorado low, forms on the east side of the Rocky Mountains in Alberta, Canada, and sweeps southeastward across the upper Midwest and Great Lakes region. A clipper always drags cold air behind it, plunging temperatures below zero across the northern Great Plains and Midwest.

Black Blizzards

As settlers moved to the Great Plains and Midwest, they tilled under the native grasses as they tamed land for agricultural pursuits and made high profits from the rich soil during times of favorable weather. The climate and weather in the plains, however, is anything but stable. The rainy periods the farmers came to expect didn't last long. Starting in 1930, a severe drought gripped the region. One of the worst, most widespread droughts in the nation's history, it lasted for ten years. As the earth dried, crops couldn't grow, leaving the bare, dusty soil vulnerable to even a breath of wind. Ignorance, greed, abusive farming techniques, and drought caused the disastrous Dust Bowl, and farmers of the plains went from feast to famine in a matter of years.

Along with limited rainfall, anomalously high temperatures (over 115°F in some places) turned the fertile plains into a virtual desert. In parts of western Kansas, drought had completely dried the soil as far as 3 feet below the surface. High winds caused by strong Colorado lows (see previous section) and cold fronts swept across the Great Plains, lifting dry soil into the air and creating massive blizzards of dust. These massive dust storms, called Black Blizzards and Blue Rollers, blew across the plains, blotting out the sun and suffocating livestock and people.

On Black Sunday, April 14, 1935, after weeks of dust storms, including one that destroyed 5 million acres of wheat at the end of March, people grateful to see the sun went outside to do chores, go to church, and enjoy the blue sky for a change. By mid-afternoon, however, the temperature dropped and birds began chattering nervously. A huge black cloud moving at 60 miles per hour appeared on the horizon. This particular Black Blizzard, considered the worst to strike the Great Plains during the 1930s, quickly dropped visibility to zero and blew loose soil into drifts several feet high. After witnessing the aftermath of the storm, Robert Geiger, an Associated Press reporter, coined the term *Dust Bowl.* This term came to include the expanse of the agricultural Great Plains—southeastern Colorado, northeastern New Mexico, and the panhandles of Oklahoma and Texas—all of which suffered through these disastrous dust storms.

Snusters

Coined in 1938, the term *snuster* refers to an awful mixture of blowing snow and dust that reaches blizzardlike conditions. Snuster storms, which occurred often during 1938 and for years thereafter, caused a tremendous amount of damage and suffering. Muddy snow plastered everything, including vulnerable livestock in open fields, and made streams and rivers murky. Snusters occur less commonly today because better farming techniques prevent loose, exposed soils from being lifted by gusty blizzard winds.

One of the worst snusters to hit the central High Plains occurred on March 31, 1955. A strong Colorado low (see *Colorado Low* in this chapter) formed and moved eastward along Colorado's southern border. The combination of the storm's moisture and the region's drought conditions caused what locals called a "mud storm." A Weather Bureau employee noted that mud had plastered all weather stations in southeastern Colorado and even the thermometers inside of the weather shelters! Winds at Las Animas and La Junta, Colorado, ranged from 40 to 50 miles per hour, with gusts up to 65 miles per hour and zero visibility for several hours. Doherty Ranch, 25 miles east-northeast of Trinidad, Colorado, reported 9 inches of muddy snow.

Massive and frightening dust storms swirled up to 8,000 feet high and made the days as dark as night. As a wall of dust loomed on the horizon, residents scrambled to stuff clothes into windowsills and in doorjambs to keep dust out of their homes. Many tried to filter the dust by breathing through wet shirts and towels. Very little could keep it out, and many people suffered from "dust pneumonia." Fatigued birds trying to out-fly the storm became engulfed in the billowing dust clouds and died. Static electricity caused by millions of dirt particles rubbing together created electrical problems for automobiles, jammed radio broadcasts, and even created balls of electricity that danced along barbed wire fences. These turbulent dust particles produced lightning and thunder. Dust storms rolled across the plains at speeds of 40 to 60 miles per hour and days later dropped heartland soil on the East Coast. Typical dust storms lasted a day or two.

A number of regional-sized (large enough to occur at more than one weather station) dust storms in the Great Plains during the 1930s dropped visibility to less than 1 mile. Weather bureau stations reported 179 smaller dust storms in April 1933 alone! Today, dust storms of this size and magnitude rarely hit.

Dust Storms of the 1930s

YEAR	OCCURRENCE
1933	38
1934	22
1935	40
1936	68
1937	72
1938	61
1939	30

Source: Soil Conservation Service

Black Blizzards left drifts of soil close to 25 feet high, and sometimes the drifts covered homes. The dust caused serious lung damage, but worst of all, these storms removed the most valuable resource of the plains—the topsoil. The Soil Conservation Service estimated that approximately 43 percent of a 16-million-acre area of the Great Plains received serious damage from these storms by the end of the 30s. The most seriously damaged area encompassed southeastern Colorado, the western half of Kansas, the Texas and Oklahoma Panhandles, and northeastern New Mexico.

The Black Blizzards made life unbearable for many in eastern Colorado and western Kansas. By 1936, more than 2 million farmers on the Great Plains and in the Midwest drew relief checks from the federal government, and as many as 3.5 million people packed up their belongings and fled west.

Farming methods have changed radically since the Dirty Thirties. During the Dust Bowl, farmers cleared the ground while harvesting and left bare topsoil exposed to the wind. Today however, farmers practice "trash farming." They leave pieces of wheat or stalks on or in the soil to hold

Top: *A Black Blizzard approaches Stratford, Texas, on April 18, 1935.* —Photo courtesy of the National Oceanic and Atmospheric Administration, George E. Marsh Album

Bottom: *Great Plains farm equipment buried in dust in 1935* —Photo courtesy of the National Oceanic and Atmospheric Administration/ Department of Commerce

The Little Dust Bowl

Just as memories of the Dust Bowl began to fade, another drought began in 1950 and peaked in 1956. This period of very dry weather became known as the "Little Dust Bowl." The 1950s drought encompassed a larger area of the United States than the 1930s drought, covering much of the Great Plains and stretching into the Southwest. While conditions were actually drier during this drought than the one in the 1930s, better farming practices reduced the amount of horrific Black Blizzards.

The most destructive Black Blizzard of the Little Dust Bowl occurred on March 10, 1955. Winds associated with the cold front of a Colorado low spread dust, sand, rain, snow, and hail over a 500,000-square-mile area, including parts of New Mexico, Colorado, Kansas, and Oklahoma. Blowing dust reduced the visibility to zero at Sedgwick, Colorado. The wind blew out windows, sandblasted cars, destroyed crops, and toppled antennas.

the soil in place and trap moisture after a harvest. Farmers also let fields lie fallow for a season, allowing precipitation to replenish the soil. Also, a post–Dust Bowl government plan called the Conservation Reserve Program paid farmers to set aside about 10 percent of their erodible acres, allowing them to revert to natural grassland.

Blue Northers: Texas Panhandle

The wind blows free in Texas. With no mountains and very little vegetation to stop it or even slow it down, the wind never seems to stop. The Great Plains—particularly the Texas Panhandle—claims notoriety as the windiest region of the country. Surprisingly, the Windy City, Chicago, has an average wind of 10.3 miles per hour, much less than the mean annual wind speed in Amarillo, Texas, which reaches 13.5 miles per hour. Amarillo, the second windiest U.S. city follows behind Dodge City, Kansas, where the average wind speed blows at 14 miles per hour.

Texas winds blow predominantly out of the south and southwest as warm air moves to the north—part of the earth's global circulation

Gimme' Shelter?

In the 1930s, prolonged winds in the Great Plains drove settlers out of their minds. Horace Greeley, a famous *New York Tribune* editor and reporter who reported on the state of new settlements for the newspaper, thought of the wind as "a confounded and constant nuisance." Oscar Wilde remarked, "The wind is a great aggravation." General Custer, however, was slightly more diplomatic: "The wind is a steady companion and bother."

Although the wind of the Great Plains was a constant bother, settlers found the territory ideal for farming, and innovative settlers decided to tame the wind by creating shelterbelts rather than giving

in to its persistence. Farmers planted windbreaks—lines of trees to reduce wind speed near the ground. These shelterbelts minimize soil erosion and wind damage to new crops and typically consist of long, narrow stands of trees planted perpendicular to the prevailing winds. During winter, shelterbelts make effective snow fences, trapping large amounts of valuable insulating snow that would otherwise blow away. The extra snow adds moisture to the fields in spring, helping to keep the soil moist and making it less susceptible to wind. Shelterbelts can protect against everyday prevailing winds but not strong winds associated with Blue Northers.

system—as air masses seek to equalize their temperatures. During spring, the windiest time of year, the average wind speed reaches nearly 16 miles per hour in the central and southern Great Plains (western Texas, Oklahoma, eastern Colorado, Kansas, and Nebraska). Winter and spring deliver extreme wind gusts associated with Blue Northers to these same areas.

Blue Northers, also known as Texas Northers, are unpleasant winds caused by massive floods of polar air. These winds sweep down from the north across Texas and Oklahoma. The high-pressure system that spawns Blue Northers forms near the north pole as bitterly cold air masses accumulate. A dome of high pressure builds, leading to light winds, clear skies, and successively colder nights in Canada and the northern Great Plains. As the high-pressure area grows, its cold, dense air spreads south, inducing a surface high-pressure area over the northern Great Plains. Intense Blue Northers need a north-south-flowing jet stream to push air rapidly south. With the jet stream positioned this way, the leading edge of the air

mass—the cold front—can race south at speeds of 40 miles per hour or more, fast enough to move it from Canada to the Gulf of Mexico in less than two days. A "norte" occurs when the cold air plunges through Texas and Oklahoma into Central America.

As its name implies, a Blue Norther event delivers a dark, bluish black sky. Its strong winds can push thick, ominous, bluish gray stratus clouds—low, sheetlike clouds—ahead of it. These winds grow more severe when they arrive in spring or fall after a period of balmy weather. Arctic winds can force temperatures from balmy 70s to the 30s in only a few hours. Temperatures can drop as much as 40°F within a few minutes and up to 60°F in a twelve-hour period. Without tall mountains between them and the Northwest Territories, the Texas and Oklahoma Panhandles often sit in the direct path of these Blue Northers as they move unhindered across the plains.

Blue Northers occur most often in winter and spring because the strong polar jet stream has its most acute meridional flow—north to

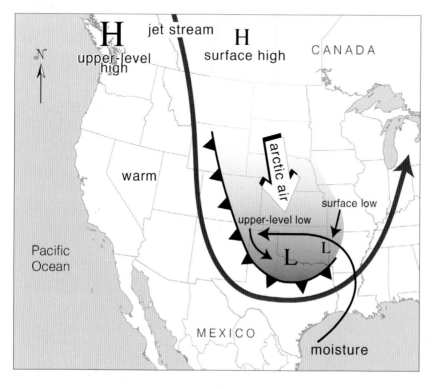

Generalized weather pattern of a Blue Norther. A ridge of high pressure develops over the Gulf of Alaska, and cool air pools near the north pole. In response a low-pressure trough develops over the Great Plains. A cold front moves south out of Canada—the Blue Norther—and over the Great Plains, bringing a lot of cold air with it. Often dry, as the cold front pushes the surface low south, it taps moisture from the Gulf of Mexico. Eventually the cold air undercuts the low, and the counterclockwise flow around the low rotates moisture over the cold air trapped near the ground. This results in heavy snow, ice, and sometimes blizzards over the central and southern Great Plains.

Notable Blue Norther–Induced Weather in the Texas and Oklahoma Panhandles

February 11–13, 1899: The worst norther on record in Texas pushed Panhandle temperatures to –23°F, left a thin sheet of ice on Galveston Bay, froze citrus groves to 12°F, and chilled people as far south as El Salvador.

February 1–15, 1905: Average daily highs dropped to 21°F and lows to near 4°F in Amarillo, Texas, with the coldest temperatures dropping into the –10 to –15°F range; normal February daytime highs in Amarillo reach the low 50s, with lows in the upper 20s.

November 11, 1911: In a matter of hours, after a Blue Norther's cold front passed, temperatures in Oklahoma City dropped from a balmy record daily high of 83°F to a record daily low of 17°F. Many areas in Oklahoma and Texas saw the temperature plummet 50°F in one hour. The front produced severe thunderstorms and tornadoes across the upper Mississippi Valley, a blizzard in the Ohio Valley and the upper Midwest, and a dust storm in Oklahoma.

March 25–26, 1934: A blizzard dumped 20.6 inches of snow on Amarillo, Texas, in just twenty-three hours. On average Amarillo receives 15 inches of snow a year.

April 6–8, 1938: Known by locals as the "84-Hour Blizzard," this storm rocked the eastern half of the Texas and Oklahoma Panhandles. This region experienced 10- to 20-foot snowdrifts, strong winds (77 miles per hour at Pampa, Texas), and zero visibility for much of the storm.

November 23–25, 1940: The Great Ice Storm of 1940 dumped freezing rain across Texas and Oklahoma for two and a half days, coating everything with 2 inches of ice while downing trees and power lines. In 1940, the National Weather Service considered it the worst ice storm the nation had seen. Power and water remained out for three days.

February 1–8, 1956: The largest unofficial snow totals occurred in the Texas and Oklahoma Panhandles: 43 inches in Vega, Texas; 24 inches at Hereford, Texas; and 14 inches in Amarillo. Snow fell for unusually long periods of time for this region, up to ninety-two consecutive hours. Twenty-three people died, as did hundreds of cows.

March 23–25, 1957: For three days Amarillo endured its most severe blizzard ever as 40- to 50-mile-per-hour winds whipped 11.1 inches of snow into drifts as high as 30 feet. The Amarillo National Weather Service office ranks this event as the third most significant weather event to affect the Texas and Oklahoma Panhandles in the twentieth century.

February 2–5, 1964: A blizzard dropped 17.5 inches of snow in Amarillo, Texas, while up to 26 inches buried Borger, Texas; normal annual snowfall for Amarillo reaches only 15 inches.

February 21–22, 1971, and March 23, 1987: Spring blizzards dropped 1 to 2 feet of snow across the Texas and Oklahoma Panhandles. Drifts exceeded 12 feet in places and stranded thousands of motorists.

December 29, 1978–January 11, 1979: Average daily highs dropped to 20°F and lows to 5°F at Amarillo, Texas. The coldest temperature, with windchill, was –43°F at Amarillo. The daytime temperatures in the Texas and Oklahoma Panhandles just prior to the Blue Norther had been in the low 60s.

November 24, 1992: Whiteout conditions near Amarillo caused a two-hundred-car accident on Interstate 40. Miraculously, no fatalities or injuries happened; however, other accidents caused by the blizzard claimed three lives.

January 28–30, 1999: A winter storm in the northern Texas Panhandle halted all travel, caused $8.5 million in damage, and knocked out power to 50,000 homes.

south—at this time of year. A Blue Norther can spark prolonged cold spells (up to a week) and sometimes blizzards across the northern half of Texas. Usually at this time of year, the air mass over northern Texas is relatively mild due to the proximity of the Gulf of Mexico, therefore preventing snow. When a low-pressure area tracks across Texas in association with a Blue Norther, however, the right mix of moisture from the Gulf and cold air from the north meet to cause heavy snow and ice.

As the arctic air moves southward, so does the high-pressure system, creating a pressure gradient and strong northerly to northeasterly winds behind the front. The dense, cold air of the southward-moving cold front will undercut the warm, moist air near the surface. This creates a shallow layer of cold air near the surface. Often the cold front will slow or stall over southern Texas, bringing an extended period of freezing rain or sleet. The cold front slows or stalls because of the increased resistance it encounters from the warm tropical air across the Gulf of Mexico. Also, by this time the cold air has mixed with the warmer marine air and has become very shallow. This is not to say that the cold fronts completely stop over the Gulf. In fact, they sometimes reach—with greatly moderated temperatures—as far south as Guatemala.

The blast of frigid arctic air can create blinding snow or ice storms. A Blue Norther can sweep in so quickly and unexpectedly that it traps and kills unprepared travelers, hunters, or other outdoor recreationists from exposure and hypothermia.

The Marfa Dryline: Western Texas

If you travel through western Texas in the summertime you may hear the term *dryline* from locals and weather forecasters. As its name implies, a dryline borders a hot, dry air mass and a warm, moist one. Drylines often occur across the southern Great Plains, particularly in western

Texas and eastern New Mexico. The dryness of the extreme western Texas and eastern New Mexico landscape, and its proximity to moist air from the Gulf of Mexico, makes Texas an ideal location for drylines. The boundary of this Texas dryline often develops near Marfa, hence its name *Marfa dryline* or *Marfa front.* During spring and summer, weather forecasters carefully watch for the Marfa dryline since drylines spawn damaging thunderstorms and even tornadoes.

Differences in terrain, climate, and vegetation help create this unique dryline that occurs on a daily basis during summer. New Mexico, dry and sparsely vegetated, has higher elevations, whereas the southern Gulf Coast of Texas has more moisture, an abundance of grasses and trees, and lower elevations. If you traveled northwest from sea level near the Gulf of Mexico to Amarillo, Texas, you would gain 3,676 feet in elevation; if you continued west to Moriarty, New Mexico, just east of Albuquerque, you would find yourself

The Lazbuddie Landspout Outbreak

Seeing a single tornado is rare, but imagine seeing six dance around you at the same time! You could have at Lazbuddie, Texas, on June 4, 1995. Six weak tornadoes, known as *landspouts,* developed and touched down within a couple square miles of a ranch outside of town. This event did not cause any damage or injuries.

Landspouts, intense vortexes, occur beneath strong, moist thunderstorms. Think of them as strong dust devils associated with thunderstorms. A *gustnado* or *gustinado,* a similar vortex, is officially considered a tornado. Weak and often short-lived, gustnados occur along the gust front of a thunderstorm. A gust front is the leading edge of gusty surface winds caused by the cool downdraft of a thunderstorm. Gustnados are often only visible as a debris cloud or dust whirl near the ground.

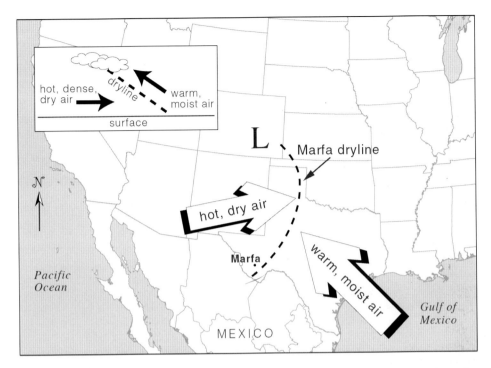

Dynamics of the Marfa dryline

hot, dense, dry air → dryline ← warm, moist air

surface

L

Marfa dryline

hot, dry air

Marfa

warm, moist air

Pacific Ocean

MEXICO

Gulf of Mexico

at 6,201 feet. This gradual increase in elevation plays a major role in the dryline's formation, since the air over the higher terrain is initially cooler (but warms dramatically during the day) and drier than that of south Texas's atmosphere.

Moist air over the lower elevations of Texas forms low, moisture-laden clouds that block morning sunshine and keep the lower elevations relatively cool, while clear skies in New Mexico allow for strong surface heating. As the sparsely vegetated topography of New Mexico warms, it vigorously heats the atmosphere above it. Two vastly different air masses result: a moist and warm one over Texas and a hot and dry one over New Mexico. This is the first step in the formation of the Marfa dryline.

The dryline often commences along the Caprock Escarpment, a prominent, rocky, north-south-trending cliff stretching from the panhandle of Texas into central Texas. It divides Texas's High Plains (the southern extension of the High Plains known as the Llano Estacado) from the lower-elevation Central Plains running east and south

of the escarpment. The escarpment rises abruptly, as much as 1,000 feet in places. Warm, moist air banks up against its east side, while dry, hot daytime air from New Mexico resides over the western side of the escarpment, known locally as the "Staked Plains." Aligned with the escarpment, the boundary between the two different air masses (the dryline) is often a focal point for thunderstorm development. To locals and storm chasers alike, the escarpment is known as the "caprock."

On spring and summer mornings, southerly winds dominate both air masses. By afternoon, though, the winds turn southwesterly across New Mexico because of the counterclockwise rotation of a low-pressure area—a heat low or stronger Pacific low—over the central Rocky Mountains. The most active drylines occur when an upper-level Pacific low-pressure system passes through New Mexico. Otherwise, on most days, a thermal low causes the wind to shift. As the strong, dry, southwesterly winds blow across New Mexico, they collide with the weaker, humid, southerly winds blowing across Texas from the Gulf. Since

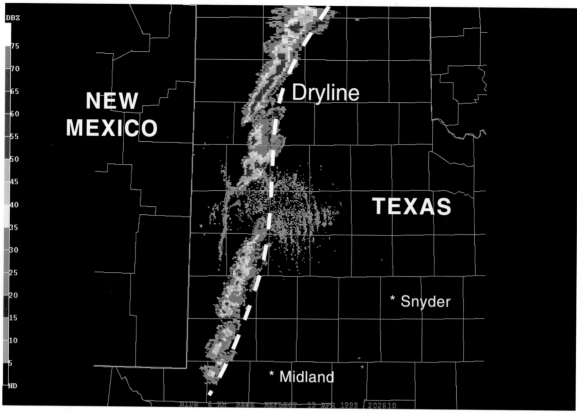

Radar image of Marfa dryline taken from the Lubbock, Texas, radar site on April 13, 1999. The line of severe storms later produced tornadoes in Midland and Snyder, Texas. —Image courtesy of the National Weather Service

A severe thunderstorm associated with a squall line in western Texas —© Weatherstock/Warren Faidley

dry air is more dense than moist air, the air from New Mexico undercuts the warm, moist air of Texas and forces it to rise and cool, forming precipitation-producing clouds. It doesn't cut very deeply, though. The once broad boundary between the air masses—where the temperature and moisture differences had ample room to coexist—is squeezed by the winds, often near Marfa, Texas. This compression creates a distinct boundary called the *convergence line*—a zone where two different air masses meet.

Dew points, the measure of moisture content in the air, can reach the 60s east of the dryline but drop off sharply into the 20s just a few miles west of the dryline. Because of these differences, the dryline is also called a *dew point front,* and easily triggered, severe thunderstorms can occur here.

By late afternoon a deep layer of hot, dry air exists over eastern New Mexico, and the warm, southwesterly New Mexico winds sometimes turn westerly. Since this western wind emanates from higher elevations—3,000 to 4,000 feet above sea level—across central New Mexico, it often remains at that elevation as it blows across Texas. A warm layer of air several thousand feet above Texas develops. This layer, known as a "cap," acts as a lid (inversion) and prevents the relatively cooler air beneath it from rising. With this atmospheric stratification in place, the westerly winds eventually push—much like a typical cold front—the dryline eastward across Texas, and force more moist air upward. The cap exists directly above and slightly east of the dryline. Gusty winds associated with the passage of a dryline often cause areas of blowing dust and sand, sometimes greatly reducing the visibility and slowing transportation.

Once the rising moist air reaches the cap it stops rising but continues to condense, releasing latent heat—stored heat energy. This latent heat eventually warms the low-level air to the temperature of the cap. As the warming takes place, instability builds, and a bubble of warm air grows beneath the cold air in the atmosphere above the cap. Eventually, the cap breaks, resulting in a powerful updraft as the built-up instability is quickly released. The updraft—the bubble of warm, moist air—rises rapidly through the cold upper atmosphere. This explosive convection forms dangerous thunderstorms—the higher and faster a warm air mass travels in cold air, the stronger the thunderstorm that forms.

Along the Marfa dryline, thunderheads can grow to heights of 50,000 feet or higher in minutes as moist air rockets skyward at 100 miles per hour. Typical mid-latitude thunderstorms have cloud tops at 30,000 to 40,000 feet; commercial aircrafts typically fly at around 35,000 feet. The resulting line of severe thunderstorms contains large hail, torrential rain, deadly lightning, and strong winds. The most powerful dryline thunderstorms are isolated from others by tens of miles. These storm cells don't have to compete for surrounding winds and moisture, and they can collect immense amounts of power.

By sundown the eastward-moving dryline slows down and begins to retreat back into New Mexico. At this point the daytime heating has ceased in New Mexico, relaxing the warm westerly wind. Usually by midnight the boundary between the two air masses mixes and becomes undecipherable. By midmorning the entire process starts over.

Heat Bursts: Oklahoma

One of the oddest and least understood atmospheric events, a heat burst occurs when a sudden rush of warm air spreads across the landscape and forces the temperature to soar—as high as 100°F or more—in the middle of the night. Heat bursts occur as nocturnal thunderstorms dissipate. Also known as "warm wakes," "heat blasts," or "hot flashes," they happen more often in Oklahoma than anywhere else in the United States. Why Oklahoma? Because Oklahoma often experiences nighttime thunderstorms and also has a dense network of weather stations positioned around the

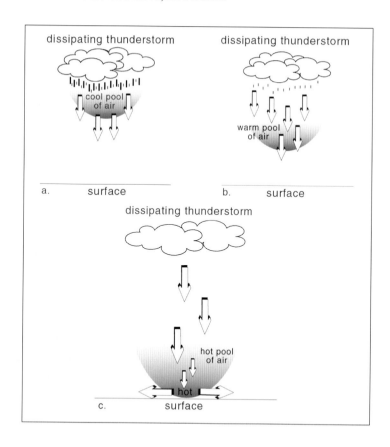

Evolution of a heat burst. (a) A pool of cool air develops beneath a thunderstorm and is held aloft, usually at night. Eventually it grows too heavy and descends from the dissipating thunderstorm. (b) As the air falls it warms through compressional heating, but it keeps its momentum. (c) As the air heats up, its downward speed slows slightly, but not enough to prevent it from slamming the earth's surface and spreading out as a heat burst.

state called the Oklahoma Mesonetwork. With as many as four in each of the state's counties, these stations can detect rare and localized events.

Under normal circumstances, a daytime thunderstorm cools the air at the earth's surface. As rain falls out of a thunderstorm, it evaporates in the dry air and cools a pool of air below the thunderstorm. The pool grows heavy and dense, and eventually it plunges toward the ground. These downbursts, which typically fall from a thunderstorm with a cloud base close to the ground (about 4,000 to 8,000 feet), consist of cool air that spreads out across the ground.

With lower surface elevations in Oklahoma than say Colorado or New Mexico, which commonly experience thunderstorms but not heat bursts, the bases of thunderstorms occur at relatively higher altitudes. The pool of air that drops from these higher storm bases warms drastically

on its descent because it has more time to heat up and higher air pressure compresses it at lower elevations. Sometimes the air travels downward over 20,000 feet before hitting the ground. As the air warms, it becomes lighter and more buoyant, which tends to slow its downward motion; however, it has sufficient momentum to slam into the ground and spread out as a hot, dry burst of wind.

The speed of the wind and its temperature change is a function of the height of the original air mass and its moisture content. Once all precipitation has evaporated from a downburst, the air actually starts warming due to compressional heating. If the air has very little precipitation to begin with, and if it begins dropping from a higher altitude, the downburst will have more time to warm before hitting the ground. These factors also affect the strength of the wind.

Heat bursts often occur during the evening. At sunset surface heating ceases, the amount of hot, dry air that fed a daytime thunderstorm decreases, and the storms usually collapse. Nocturnal thunderstorms, however, are largely driven by lines of converging air, also known as *outflow boundaries,* which are set in motion during the heating hours of the afternoon. Nocturnal thunderstorms also rise from uplift caused by cold fronts, drylines, mountains, and other nearby thunderstorms. Strong, moist, complex low-level jet streams that tend to peak at night feed these nighttime storms. Because nighttime radiational cooling has caused surface temperatures to drop, the temperature contrast between the surface temperature and the heat burst can be dramatic.

Scientists have detected and studied more heat bursts in Oklahoma than anywhere else in the world thanks to the 111 stations that make up the Oklahoma Mesonetwork. These stations collect information about temperature, humidity, surface air pressure, wind speed and direction, soil temperature, and incoming solar radiation—data used for research and forecasting techniques. A heat burst's intensity depends on the size of the collapsing storm and how much the air warms before hitting the ground. The winds accompanying a heat burst can reach damaging speeds of 100 miles per hour. Some recorded heat bursts have affected areas of 5 square miles or less.

A particularly damaging series of heat bursts occurred over a large portion of southwestern and central Oklahoma during the evening of May 22, 1996. The winds associated with the heat bursts knocked trees over, ripped down power lines, and lifted roofs off homes and businesses. Wind gusts reached 105 miles per hour in Tipton, 95 in Lawton, 67 in Ninnekah, and 63 in Chickasha. The temperature in Chickasha, just southwest of Oklahoma City, jumped from 88 to 102°F degrees between 9:00 and 9:25 PM.

On April 26, 1974, a heat burst at Midland, Texas, drove the temperature up 46°F within a few hours. Perhaps the world's most phenomenal heat burst ever recorded occurred at Wichita, Kansas, on May 28, 1974, when the temperature jumped 52.2°F in just 20 minutes—that's a temperature rise of 2.6°F per minute! On the night of June 15, 1960, the hottest heat burst ever documented in the United States struck northwest of Waco, Texas, at Kopperl. Winds gusted over 80 miles per hour as the temperature rose to an estimated 140°F, wilting and killing vegetation. More recently a heat burst at San Angelo, Texas, drove the temperature from 70°F at midnight to 95°F at 1:05 AM on August 14, 2002.

Heat bursts can occur anywhere thunderstorms roam, even as far north as Montana. A heat burst in Glasgow, Montana, on September 9, 1994, drove the temperature from 67°F to a record daily high temperature (for Glasgow) of 93°F in just fifteen minutes. In Greensburg, Kansas, the temperature on July 7, 1987, rose from 75°F at 7:00 AM to 95°F in a few minutes. Dust devils and what an observer described as "strange clouds" accompanied this event.

Infrequent and mysterious, heat bursts only affect localized areas, but they can have a staggering effect on temperature, not to mention their winds, which can do serious damage. Any dying thunderstorm with a high base can produce a heat burst if enough evaporation has occurred before the pool of cool air reaches the ground. Though relatively unusual, no one knows how many heat bursts occur because weather instruments often don't detect them.

Tornado Alley

The strongest and most deadly tornadoes in the world frequently strike the midsection of the United States, which is known as Tornado Alley or the Tornado Belt. Tornado Alley, while loosely defined, is a region that stretches across Texas, Oklahoma, Kansas, eastern Colorado, eastern Nebraska, Iowa, Missouri, western Arkansas, and western Louisiana. A unique mix of topography

A tornado rages 8 miles northeast of Stapleton International Airport in Denver, Colorado, on May 18, 1975. This 1,320-foot-wide tornado lasted twenty minutes and traveled 4½ miles in open fields. It caused no damage.
—Carol A. Silberg photo, courtesy of the National Oceanic and Atmospheric Administration/Department of Commerce

The West's Strongest Twister

The scientific community widely accepts that the strongest tornado to strike west of the hundredth meridian touched down just east of Thurman, Colorado (21 miles north of Flagler), on August 10, 1924. The twister killed eleven people in a single home. Residents of Thurman described it as "huge" and "cone-shaped." Post-storm investigation meteorologists estimated the winds blew at 260 miles per hour. The twister likely ranked as an F4 or F5 tornado.

and clashing air masses makes Tornado Alley the most tornado-stricken region of the world.

Geographical location, a supply of moist air, and wind shear make this region of the United States susceptible to tornadoes. First, not as many tornadoes would occur in the Tornado Belt but for the Rocky Mountains to the west, the Gulf of Mexico to the south, the deserts of northern Mexico to the southwest, and the gently sloping Great Plains. The topography and associated weather patterns of all these regions work together to spawn tornadoes.

Second, warm and humid surface air under cooler, drier air produces an unstable atmosphere in the Tornado Belt in the spring and summer. The cooler, drier, upper atmospheric air blows over the Rocky Mountains. The Gulf of Mexico provides an abundant supply of tropical moisture, which blows into the plains as southerly and southeasterly winds.

The third factor, vertical wind-shear—change in wind speed or direction with altitude—plays a key role in the development of rotating thunderstorms called *supercells* or *mesocyclones*. Unlike typical thunderstorms, made up of a single updraft and a single downdraft, supercells are dangerous convective storms that consist of a single, powerful rotating updraft, which persists for a long period of time. Supercells spawn the biggest tornadoes, but to form they need low-level and mid-level winds (surface to 20,000 feet) shifting direction and growing stronger with height. These wind conditions can easily develop over Tornado Alley when a trough of low pressure—attached to the prevailing westerly jet stream—moves eastward across the Great Plains. While the trough moves, cool high-level westerly winds (30,000 feet and higher) blow from the Pacific Ocean and Rocky Mountains.

At the same time warm, dry, southwesterly winds blow from the deserts of New Mexico and Mexico at mid levels of the atmosphere (5,000 to 20,000 feet). Because these winds emanate from higher elevations, they often remain at that elevation as they blow across Texas. This results in a warm layer of air, usually at the lowest level of the mid-level air mass, several thousand feet above Tornado Alley (the same pattern that helps create the Marfa dryline). This layer, or "cap," acts as a lid (inversion), preventing the relatively cooler air beneath the cap from rising.

And lastly, at the low levels of the atmosphere (less than 5,000 feet), moist southerly air blows into the region from the Gulf of Mexico, often in the form of a low-level jet stream. This moisture-rich jet stream is critical because it provides the moisture and low-level wind shear that is necessary for supercells to form. In other words, the three levels of the atmosphere have different wind directions and temperatures: cold upper-level winds from the west, warm mid-level winds from the southwest, and relatively moist, cool low-level winds from the south.

When the rising moist air reaches the mid-level cap, it stops rising, but its water vapor starts to condense and release latent heat. This latent heat eventually warms the air to the temperature of the cap. As the warming takes place, instability builds, and a bubble of warm air grows beneath the cold air of the atmosphere above the cap. Eventually, the cap breaks and the built-up instability escapes, resulting in a powerful updraft. The updraft—the bubble of warm, moist air—rises rapidly through the upper atmosphere. This explosive convection causes dangerous thunderstorms to form quickly—the higher and faster a warm air mass travels in cold air, the stronger the thunderstorm that forms. The differing wind directions at different elevations put a counterclockwise spin on the plume. This spinning motion, with the correct conditions, forces the entire thunderstorm to rotate, producing a rotating thunderstorm, or supercell. Once born, a highly organized supercell can easily drop a tornado from its base.

As the spinning increases due to wind shear, a mesocyclone—a rotating thunderstorm updraft—can form in the middle of the storm. Mesocyclones are usually 3 to 12 miles wide. As

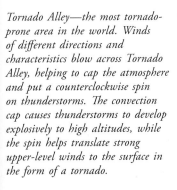

Tornado Alley—the most tornado-prone area in the world. Winds of different directions and characteristics blow across Tornado Alley, helping to cap the atmosphere and put a counterclockwise spin on thunderstorms. The convection cap causes thunderstorms to develop explosively to high altitudes, while the spin helps translate strong upper-level winds to the surface in the form of a tornado.

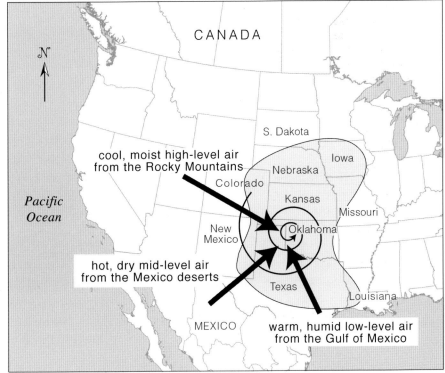

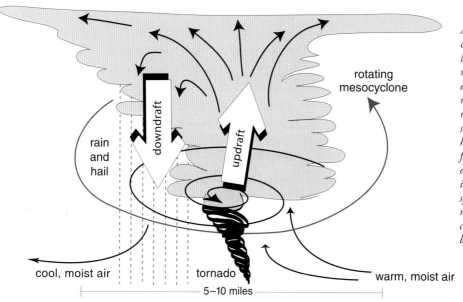

A tornadic thunderstorm. Winds of different speeds and at different levels of the atmosphere cause the mesocyclone to spin. The storm draws in vast amounts of warm, moist air that narrows into a rotating updraft. As the mesocycl[e] stretches vertically and shrinks horizontally, the updraft spins faster, much like an ice skater wh[o] draws her arms toward her body in order to spin faster. The updra[ft] spins into a fast, tight vortex in [the] mid levels of the mesocyclone. It can then extend to the ground a[nd] become a tornado.

air rushes into the low-pressure area the meso-cyclone creates, the strong updraft can force it to narrow. Like a spinning skater who increases her speed as she draws her arms toward her body (the conservation of angular momentum), the mesocyclone gains speed as it stretches verti-cally and shrinks horizontally. Meteorologists call this process *stretching,* and it greatly increases a mesocyclone's circulation.

The mesocyclone, 1 to 2.5 miles wide after stretching, continues to spin faster until a spin-ning vortex forms in the mid levels of the storm. This spinning vortex gradually extends down-ward and becomes a "wall cloud," an abrupt lowering from the base of a cumulonimbus cloud. It looks like a low-hanging cloud, and it normally measures ½ mile or more in diameter. A wall cloud marks the lower portion of a super-cell's very strong updraft. If wall clouds exhibit rotation, they may spawn a tornado within a few minutes to one hour. Sometimes the de-velopment of a wall cloud is the only warning that a tornado is coming. Sometimes large hail precedes a tornado as well.

If the stretching process continues, a narrow funnel may result from a wall cloud; meteorolo-gists call this a *funnel cloud.* As air near the surface is drawn into the very low pressure of the mesocy-clone, adiabatic expansion causes cooling, inducing condensation and creating cloud droplets. Cloud droplets make the funnel visible, and meteorolo-gists call this visible vortex a *condensation tube.* As this process keeps up and the vortex continues stretching, the funnel will extend to the ground and become a tornado. Sometimes dust and debris will stir at the surface before a condensation fun-nel reaches the ground, but once the condensation tube reaches the ground it violently sucks in addi-tional dirt and debris, making the tornado appear dark and ominous. With the aid of sophisticated weather radar, scientists have discovered that large, violent tornadoes have smaller whirls, known as *suction vortexes,* rotating within a parent tornado.

In Tornado Alley, tornadoes occur most frequently during spring. About three-fourths of all tornadoes in the United States occur between March and July, peaking between May and June. The most violent and deadly tornadoes tend to occur in early spring and even late winter when strong winter winds continue to stir the atmosphere. At these times of year the differences in speed and direction between the colliding air masses are most striking, so they collide and mix with explosive results. Plus, upper-level winds are stronger as potent spring storms sweep across the nation with the mighty jet stream that blows over 200 miles per hour. The jet stream provides the wind shear between the mid and upper levels of the atmosphere, and the faster the jet stream moves the more intense the shear.

Measuring a Tornado's Fury

Meteorologists use the Fujita Scale—a system devised by and named after the famous tornado expert Dr. T. Theodore Fujita—to categorize tornadoes. The amount of a tornado's damage determines its rank on the scale.

One of the most damaging Colorado tornadoes hit Limon in 1990. At 8:08 PM on June 6, an F3 tornado tore apart the sleepy town of Limon.

The violent twister demolished 117 homes and 23 businesses, leaving 155 people homeless, and inflicting more than $12.4 million ($17.1 million in 2002 dollars) in damage. The twister raged across the business district and the railroad yard, flipping 30-ton hopper cars, tearing apart silos, and leveling 80 percent of Limon. Fourteen people sustained injuries, but miraculously all survived since the National Weather Service provided the community sufficient advance warning.

Texas, in part due to its size, has recorded more tornadoes (6,633 between 1950 and 1999) than any other state in Tornado Alley. The United States alone endures an average 760 tornadoes a year—with up to 1,300 in some years (for example, 1992)—and 91 tornado-related deaths. In 1953 the United States experienced more tornado-related deaths than in any other single year: 519 people died, 114 of those in a twister that roared through Waco, Texas, on May 11. The National Weather Service ranks the Waco twister as the tenth deadliest U.S. tornado.

Usually tornadoes in Tornado Alley occur during the late afternoon and early evening because

Tornado F-Scale

RANK	CATEGORY	MILES PER HOUR	DAMAGE
F0	Weak	40–72	Light: some damage to chimneys, tree branches broken, and shallow-rooted trees knocked over
F1	Weak	73–112	Moderate: shingles peeled off roofs, mobile homes pushed off foundations or overturned, moving cars and trucks blown over and off roads
F2	Strong	113–157	Considerable: roofs torn off houses, mobile homes demolished, train boxcars overturned, large trees broken or uprooted, cars and small trucks lifted off the ground
F3	Strong	158–206	Severe: roofs and some walls torn off well-constructed houses, trains overturned, most trees uprooted or snapped, heavy cars and trucks lifted off the ground and thrown
F4	Violent	207–260	Devastating: well-constructed houses leveled, structures with weak foundations blown away, cars and trucks thrown
F5	Violent	261 or more	Incredible: sturdy houses leveled and swept away, large trucks and cars picked up and thrown great distances, trees debarked, asphalt roads desurfaced, and other incredible phenomena occur (for instance, straws of hay driven into telephone poles)

the earth's surface warms during the day, but they also occur at other times. Most tornado deaths occur at night or in areas with thick vegetation, because people don't see them coming. Perhaps the deadliest place during a tornado is inside a vehicle. If you see a twister while driving, the National Weather Service advises you to stop immediately and seek shelter in the interior of a building or a low-lying ditch. Even though Tornado Alley is the most tornado-prone area of the world, if you stood in the middle of Oklahoma—the center of Tornado Alley—you would have to wait an average 1,400 years before experiencing a direct hit.

Top Ten Deadliest U.S. Tornadoes

RANK	STATES	DATE	DEATHS	INJURIES	F-SCALE CATEGORY
1.	Missouri, Illinois, and Indiana	March 18, 1925	695	2,027	F5
2.	Louisiana and Mississippi	May 7, 1840	317	109	Unknown
3.	Missouri and Illinois	May 27, 1896	255	1,000	F4
4.	Mississippi	April 5, 1936	216	700	F5
5.	Georgia	April 6, 1936	203	1,600	F4
6.	Texas, Oklahoma, and Kansas	April 9, 1947	181	970	F5
7.	Louisiana and Mississippi	April 24, 1908	140	770	F4
8.	Wisconsin	June 12, 1899	117	200	F5
9.	Minnesota	June 8, 1953	115	844	F5
10.	Texas	May 11, 1953	114	594	F5

Tornado damage in Moore, Oklahoma, after a violent F3 tornado plowed through it and Oklahoma City on May 8, 2003 —Bob McMillan photo, Federal Emergency Management Agency

A tornado in the Oklahoma Panhandle on May 5, 1993 —©Weatherstock/Warren Faidley

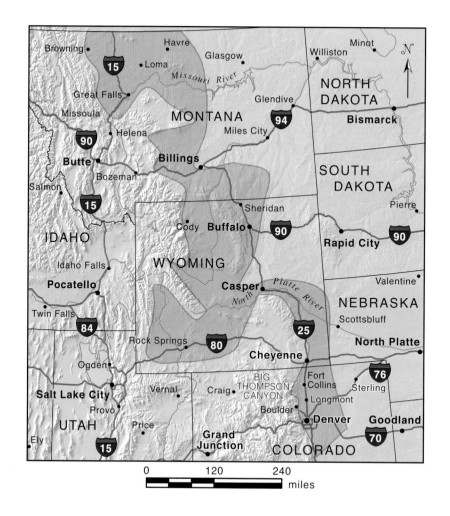

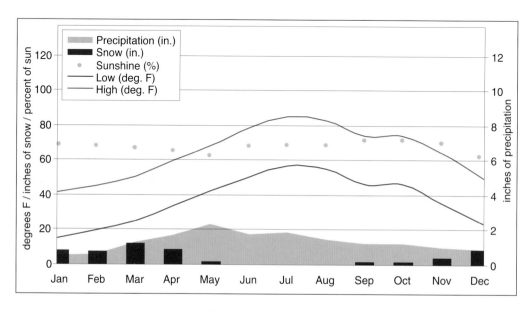

Climate of Denver, Colorado

12

Rocky Mountain Front and Adjacent Foothills

Elevation: 2,500 to 9,500 feet

Natural vegetation: grass and other herbaceous plants, and scattered needleleaf evergreen trees

Köppen climate classification: dry, particularly in the summer, with cool temperatures

Most dominant climate feature: strong winds off the Rocky Mountains in fall, winter, and spring

Strong wind and cold temperatures dominate weather features during the months of January and February along the Front Range. Although often warm, westerly winds blowing off the Rocky Mountains can blow violently, sometimes exceeding 100 miles per hour along the foothills. Temperatures during these two months can range wildly—from the 50s to the teens—in a single day, but average daytime highs reach the 20s in the north in Montana and the 40s in the south in Colorado. Morning lows range mostly between 0 and 20°F. Light precipitation falls, and less than 1 inch (including melted snow) accumulates across the region.

March and April, among the most active weather months of the year for this region, might deliver anything from a heat wave to a major blizzard. What snow does fall, however, usually melts within a day or two since warm temperatures and clear skies replace the wintry weather. In March daytime temperatures warm into the 30s in the

north and the 50s in the south. Daytime temperatures reach the 60s by April, with nighttime lows in the 20s and 30s. Normally, 1 to 3 inches of precipitation, falling mostly as snow, accumulates during this two-month period. The driest conditions occur in the north because the storm track tends to travel well south of Montana at this time of year.

The last spring freeze usually occurs in May, but northern areas can have their last freeze in June. Temperatures continue to rise in May and June with daytime highs reaching the 70s and overnight lows in the 40s and occasionally the 50s. The wettest month of the year, May brings large-scale storms providing upslope flow and cloudy skies throughout the area. Even a rare snowstorm can blow in at this time of year. Average precipitation for May and June combined measures 2 to locally 6 inches. May marks the beginning of hail season. Though mostly small, hail at this time of year occurs frequently with almost daily thunderstorms.

July and August are active thunderstorm months and can deliver severe storms that produce flooding rains, high winds, and sometimes tornadoes. Winds may blow in excess of 57 miles per hour and damaging hail ¾ inch in diameter or greater can fall in the south. Total precipitation for the July-August period ranges usually in the 1- to locally 4-inch range, with the wettest conditions occurring south—closer to the Pacific Ocean and

Gulf of Mexico, which provide the region with monsoonal moisture (see *Southwest Monsoon* in *chapter 6*). Temperatures can exceed 100°F, but ordinarily daytime readings warm to the 80s and 90s, with overnight lows cooling to the 50s.

In September and October, daytime highs are mostly in the 60s and 70s, but overnight lows drop into the 30s and 40s. The first fall freeze typically occurs in September when the leaves begin to change color. The first shot of wintry weather occurs in September in the north, and the first measurable snow in the south typically falls in October. For the most part, however, calm weather prevails at this time of year. Precipitation totals for the two-month period range from 1 to 3 inches.

Many areas see a peak in snowfall amounts during November, with a slight decrease in December. Major snowstorms can deliver a paralyzing blow to the region with strong winds up to 80 miles per hour, a lot of snow (up to 24 inches), and bitterly cold (subzero) temperatures. If such a storm occurs, snow can remain on the ground for weeks to a month because of cold temperatures, low sun angle, and limited daylight hours; however, a warm Chinook could arrive and easily melt the snowpack in a day or two. Precipitation for the period averages from a trace to 2 inches. December, the driest of the two months, sees temperatures range from the 30s and 40s during the day to the teens and 20s during the night.

The Denver Cyclone

With the Front Range of the Rockies to the west and the Great Plains spanning to the east, the topography of Denver, Colorado, creates a unique counterclockwise circulation of air known as the Denver Cyclone. Not until 1980, with the establishment of a dense network of automated weather stations—the Program for Regional Observing and Forecasting Services (PROFS)—in northeastern Colorado, did meteorologists recognize this complex, frequent, small-scale circulation.

The Denver Cyclone, a localized phenomenon induced by winds that curl around and over local topography, strays between Denver and Greeley. The terrain north of Denver gradually increases in elevation—1,150 feet—to the crest of the Cheyenne Ridge (6,000 feet), which extends from Cheyenne east to where Colorado, Wyoming, and Nebraska meet. More importantly, however, a ridge called the Palmer Divide climbs 2,625 feet higher than Denver and runs in an east-west direction between Denver and Colorado Springs. West of Denver, the Rockies rise to the Continental Divide, which averages about 11,150 feet of elevation in this region. The Great Plains stretch off to the east from the Front Range, sag somewhat, and leave Denver in a basin.

The Denver Cyclone is a lee eddy. Also called a *lee-side low*, this eddy is a low-pressure area that forms downwind of Palmer Ridge on days with fairly stable conditions. As south to southeasterly winds blow across southeastern Colorado and up Palmer Divide, they are squeezed into a smaller vertical layer of the atmosphere—bound by the Palmer Divide below and relatively warmer air aloft, which acts as a lid. The lid and mountains force the winds to expand horizontally, slowing them down and causing them to curve counterclockwise. As the winds blow down the north side of Palmer Ridge, the air mass once again stretches vertically. The counterclockwise spin of the air mass grows stronger, and it develops into a small-scale low-pressure area over Denver. Like an ice skater pulling in his arms to spin faster, the cyclone's strength and spin increase when it gains vertical space in which to develop. With time the circulation grows stronger in the Denver area, and it can create severe thunderstorms. The relatively broad (in comparison to a tornado), low-level spin associated with the Denver Cyclone promotes rotation in thunderstorm updrafts, increasing the potential for tornadoes.

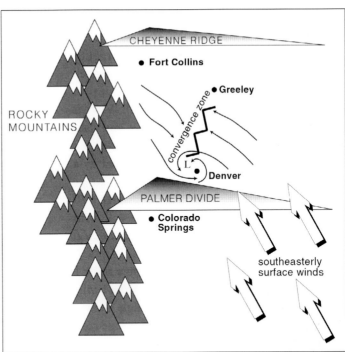

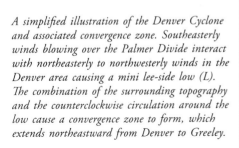

A simplified illustration of the Denver Cyclone and associated convergence zone. Southeasterly winds blowing over the Palmer Divide interact with northeasterly to northwesterly winds in the Denver area causing a mini lee-side low (L). The combination of the surrounding topography and the counterclockwise circulation around the low cause a convergence zone to form, which extends northeastward from Denver to Greeley.

Once formed, peripheral winds blowing on the north and east sides of the cyclone's circulation converge with the air outside of the circulation. These converging winds pile up, and since they have nowhere else to go, rise into the atmosphere—meteorologists call this *convergence.* This volatile zone of converging air, called the Denver Convergence Vorticity Zone (DCVZ), creates a place where bands of heavy precipitation and lines of thunderstorms form. During the afternoon, when conditions become unstable due to rapid surface heating, the Denver cyclone weakens, but not before initiating several areas of convergence northeast of Denver. These areas become the focal point for afternoon thunderstorm development.

Other times the convergence zone only creates a band of scattered clouds while the remainder of the region remains clear. The Denver Cyclone and its convergence zone have been known to meander between Denver and Greeley, placing the Interstate-25 corridor from Fort Collins to Denver under their thunderstorm-inducing

influence. They can last up to thirty-six hours, but typically they last six to twelve hours, which allows time for the ambient airflow entering Denver to adapt to the lee-side eddy and enhance its wind pattern.

Vertical wind-shear, one of the most dramatic by-products of the Denver Cyclone, is a change in wind speeds and direction with elevation. The prevailing winds aloft blow westerly and at higher speeds, while surface winds associated with the cyclone blow in a counterclockwise direction at slower speeds. When strong enough, the wind shear causes small tornadoes to form from DCVZ-induced thunderstorms. Relatively light winds associated with the Denver Cyclone impose weak spin in the atmosphere, which results in weak tornadoes known as *landspouts* or *non-supercell* tornadoes. The DCVZ is one of the most prolific landspout-makers in the world. Unlike violent supercell tornadoes, spawned from rotating thunderstorms known as *supercells* (see supercell discussion in *Tornado Alley* in *chapter 11*), the usually weak landspout has the shape of a rope.

University of Oklahoma meteorologist Howard Bluestein first introduced the term *landspouts* because of their resemblance to waterspouts. They have similar wind speeds, and nonsupercell thunderstorms spawn them, but they occur over land rather than water. The National Weather Service treats landspouts as tornadoes and usually classifies them as F0, F1, or in rare cases F2 tornadoes. Anything stronger, likely caused by a rotating, supercell thunderstorm, becomes a true tornado. Studies have shown that better than a 70 percent chance exists that a landspout will form daily somewhere in or near a well-formed DCVZ in June.

A typical Denver Cyclone can range from 1 to 3 miles wide; however, smaller (less than 1 mile across) eddies sometimes occur in a single Denver Cyclone. Not only does the Denver Cyclone contribute largely to summer weather by bringing severe thunderstorms, but also the circulation of the small cyclone sparks heavy snow and strong turbulence in the Denver area.

Damaging Denver Twisters

The Denver Cyclone gets credit for several particularly severe storms in the Denver area since 1981. One of the most damaging occurred on June 3, 1981, when two tornadoes ripped through Denver causing $15 million ($29.7 million in 2002 dollars) in damage. One of the twisters, the most damaging tornado to hit the Denver Metropolitan area, ripped through 2.3 miles of Thornton. The tornado destroyed eighty-seven homes, damaged another six hundred, and injured forty-two people—seven seriously. The other tornado occurred near the Denver-Lakewood boundary.

The Denver Cyclone caused the famous June 13, 1984, Denver hailstorm—the costliest hailstorm in U.S. history—that totaled $350 million ($605.5 million in 2002 dollars) in damage. The

DCVZ, with its clashing air masses and severe thunderstorms, provides an area where strong updrafts form. Updrafts are a key ingredient in the formation of hail. The updraft's upward-flowing air suspends frozen raindrops and tosses them throughout a storm. As the frozen raindrops rotate within the storm, they collect more ice as they pass through the warmer (above-freezing) lower portion of a thunderstorm. They grow and become hailstones. Without an updraft these raindrops would fall to the ground before growing too large. Eventually, updrafts can no longer keep the heavy hailstones suspended, and they fall to the ground—the stronger the updraft the larger the hailstones.

The Longmont Anticyclone

The interaction of surface winds and the terrain along the northern Colorado Front Range create yet another circulation feature. Like the Denver Cyclone, the Longmont Anticyclone forms a mesoscale zone (50 to 250 miles wide) of curling winds in the lee of higher terrain. Unlike the Denver Cyclone, the winds of the Longmont Anticyclone spin clockwise.

The Longmont Anticyclone forms when low-level northwesterly winds interact with the east-west-oriented Cheyenne Ridge, which lies along the northeastern Colorado–Wyoming border. The center of the anticyclone often falls just east of the foothills of the Rocky Mountains near Longmont, Colorado. As northwesterly winds flow down the south side of Cheyenne Ridge, they turn easterly across northern Colorado's Front Range, spinning in a clockwise direction; in some cases, the winds will turn southerly in places like Fort Collins. This feature sometimes enhances upslope (easterly) precipitation west of Denver.

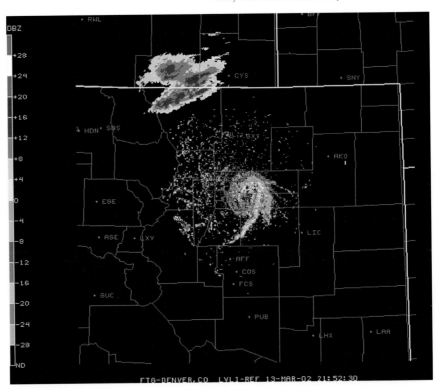

A well-developed Denver Cyclone over Denver on the afternoon of March 13, 2002. Since this Denver Cyclone event didn't drop any precipitation, the radar detected concentrations of particles, such as dust and water droplets, in the atmosphere. Greens and blues indicate lighter concentrations, while yellows and red indicate higher concentrations. Higher winds kicking up more particles likely caused the high concentrations. Image courtesy of the National Oceanic and Atmospheric Administration/National Weather Service

During winter, the Denver Cyclone produces and maintains localized bands of heavy snow north and east of Denver. Fortunately, sophisticated weather computer models, coupled with a network of weather stations, can predict the timing and location of the cyclone, therefore improving short-term weather forecasts and warnings.

Deadly Downbursts: Denver, Colorado

Its position along the Rocky Mountain Front makes Denver, Colorado, one of the West's locations most prone to downbursts. Also known as *microbursts* or *macrobursts* depending on their size, downbursts are violent, often deadly, winds. Even though downbursts can occur anywhere in the United States, Denver has experienced very dramatic ones, including several that have caused catastrophic airline crashes.

A downburst is a severe downdraft—a swift column of air—that drops from the bottom of an elevated thunderhead or cumulus cloud that usually doesn't progress into a thunderstorm. This air blasts the earth's surface with powerful winds. Downbursts hit suddenly, and often they come from the most inconspicuous looking cumulus clouds. They only become visible when dust and debris from the surface is kicked up by their winds, or when virga falls from clouds. Downbursts usually affect a small area, typically less than 6 square miles, and last less than twenty-five minutes. Although small and short-lived, downburst winds can cause severe damage, injuries, and even fatalities. The winds can gust over 100 miles per hour, topple trees, tumble cars and trains, and rip the roofs off houses.

Downbursts need a deep layer of hot, dry air near the earth's surface and a moist cloud-producing layer of air at least 5,000 feet above the surface. The higher the cloud base, the stronger the downburst potential; in the most severe downbursts, the cloud base rises more than 14,000 feet above the surface. A cumulus cloud will form in the moist layer, and the dry lower

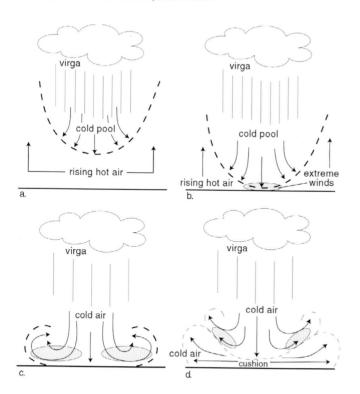

Evolution of a downburst. (a) A downburst initially develops as evaporating rain (virga) cools a pool of air below a cumulus cloud. (b) As warm rising air displaces the pool, it descends from the cloud. As it descends, it accelerates and reaches the ground within minutes. This is known as the contact stage. During the contact stage, the strongest winds rage at the surface. (c) In the outburst stage, the winds curl as the cold air is deflected from the point of impact. (d) During the cushion stage, a cold spreading layer creates a cushion for the continuing downdraft. The curling winds continue to blow outward and upward, posing an enormous threat to aircraft flying in the area.

layer of the atmosphere whips into the storm as an updraft. Since the air is dry, it cannot supply the cloud with the water vapor necessary to create a rainstorm or thunderstorm.

Given enough moisture over 5,000 feet, the cloud will grow large, climbing high enough into the atmosphere to produce rain. The rain, however, evaporates in the dry lower layer before hitting the ground, causing gray streaks in the sky called *virga*. Virga is often the first sign of a downburst. As the virga falls, the evaporative process cools the air below the cloud, and a pool of cold air develops. Set in motion by the falling raindrops, the pool of dense, cold air—now a downdraft—races to the ground at high speed (40 to 100 miles per hour, depending on the cloud's height). Two to five minutes after leaving the base of the cloud, it strikes the earth's surface, sending strong horizontal winds outward.

Moisture and heat in the lower atmosphere affect the speed of a downburst. Drier air promotes more evaporative cooling, while hot weather creates large parcels of hot air, which rise off the ground. As these large parcels rise like gigantic balloons, they displace colder air above them and deflect it toward the earth.

Ordinary downbursts easily produce winds of 30 to 40 miles per hour; however, the most severe downbursts produce sudden, dramatic wind gusts of 60 to 100 miles per hour. The air temperature at the surface can drop 10 to 20°F in a matter of seconds, fully recovering thirty to forty minutes later. A microburst can quickly turn a calm, hot moment into a dusty, windy, and cool one. Microbursts usually occur in the late afternoon when surface heating has reached its maximum.

Downbursts and their curling and downward-blowing winds of differing speeds and directions create dangerous conditions for aircraft. Nearly invisible and difficult to avoid, downbursts occur commonly throughout the world. In the United States, they have caused major aircraft mishaps in Denver, Detroit, Tucson, New Orleans, Dallas,

A downburst kicks up dust and debris near Pawnee Buttes in northeastern Colorado on September 4, 1982. A few minutes after this picture was taken, winds gusted to 57 miles per hour and the temperature dropped 11°F. —Photo courtesy of the University Corporation for Atmospheric Research

and Atlanta. Downbursts have caused about twelve major aircraft mishaps in the United States since 1975 and perhaps hundreds of minor aircraft accidents.

In a ten-second period, as a jetliner passes through a downburst, it may see a 30-mile-per-hour headwind change to a 30-mile-per-hour tailwind. Airplanes can literally get thrown to the ground. Since downbursts occur below 5,000 feet and there is no time to regain altitude before hitting the ground, the danger is especially high during takeoff and landing. Officials at Denver International Airport have sensing devices with a forty-five-second warning, enough time to warn pilots about a possible downburst. Officials often close runways when virga occurs. One of the worst downburst-caused aircraft accidents in the United States occurred at Dallas–Fort Worth Airport on August 2, 1985. A Delta 747 crashed on its final approach, killing 133 people.

Derechos

In Spanish, *derecho* means "straight." In the weather world, derechos mean extreme and widespread, convectively induced straight-line windstorms. In the family of downburst winds associated with cumulus clouds and thunderstorms, derechos can produce sustained winds of 50 to 60 miles per hour for over thirty minutes, with occasional gusts over 120 miles per hour. By definition, derechos are widespread events, affecting areas as large as 700 square miles. First recognized in the Corn Belt of the United States, they have since been observed in many other areas of the United States, including Oklahoma, New York, and Michigan.

Colorado Downbursts

August 7, 1975: During takeoff a Continental Airlines 727 was thrown to the ground by a downburst at Denver's Stapleton International Airport. Fourteen people were injured.

August 14, 1978: Dangerously fierce downburst winds gusted to 101 miles per hour at the National Center for Atmospheric Research's Mesa Laboratory in Boulder, Colorado. The strong winds caused many car accidents in the Denver-Boulder region.

July 14, 1989: A potentially deadly thunderstorm whipped up 60-mile-per-hour winds at a National Guard encampment at Pinon Canyon, 35 miles east of Trinidad, Colorado. Flying debris and downed tents injured about one hundred soldiers. Downburst winds reached 77 miles per hour at La Junta, Colorado, later the same day.

May 16, 1995: One-hundred-mile-per-hour downburst winds blasted Lamar, Colorado, and left considerable damage.

Snow Eater: Boulder, Colorado

"Snow eater," the "Boulder wind," "Chinook"—all these names describe the violent warm winds that descend the Rocky Mountains and strike adjacent foothills from New Mexico to Montana. *Chinook,* a Native American word that means "snow eater," can easily melt several inches of snow in a matter of hours. In 1986, residents of Kipp, Montana, watched 36 inches melt in less than twelve hours. The temperature rose 34°F in seven minutes and reached 80°F in a just a few hours. But more significantly, high winds of Chinooks sometimes blow faster than hurricane-strength winds.

Boulder, Colorado, northwest of Denver, falls prey to fierce downslope winds, including Chinooks, because of the surrounding topography. The Rocky Mountains rise steeply from the western edge of town. The Flatirons, a steep rock face west of Boulder that rises approximately 2,000 feet, furnish an ideal cliff for the westerly winds to race down. East of Boulder the flat terrain allows winds to flow east freely. While meteorologists commonly regard Dodge City, Kansas, as the windiest U.S. city, with an average annual wind speed upwards of 14 miles per hour, Boulder gets rocked by some of the highest wind gusts of any U.S. city. These amazing windstorms hide, however, in Boulder's low average annual wind speed of 9.4 miles per hour; nonetheless, winds of 140 miles per hour—enough to lift an average adult off the ground—have blasted Boulder.

A Chinook requires several atmospheric features. A strong jet stream must blow at about 15,000 feet from the northwest to the southeast. As with any jet stream in the northern hemisphere, it separates a colder, more unsettled air mass to the north from a warmer air mass to the south. In a Chinook situation, the jet blows between a strong, vertically stacked low—one that penetrates all levels of the atmosphere—over western Nebraska, and a strengthening high-pressure area over the Four Corners. The strongest winds, both at the surface and aloft, occur where the circulations around the high and low meet. Not only do their circulations complement each other where they meet, but the pressure gradient peaks there as well. Add the downslope component of the Rocky Mountains and the result is a fierce wind from the surface up to about 10,000 feet. Compressional heating causes the wind to warm as it descends the Front Range.

The high winds of this weather pattern can occur anywhere along the Rocky Mountain Front, from Colorado to Montana. These conditions occur reliably each year and combine to unleash an average of five Chinooks—with winds in excess of 70 miles per hour—over Boulder each year.

With everything in place, Chinooks in Boulder will commonly blow for twelve hours at

Cold and Gusty Montana Winds

Not all westerly winds along the Front Range bring warm temperatures. In the winter a downslope wind, called a *katabatic* or *bora* wind, develops when a pool of subzero air—formed by high pressure and consecutive nights of radiational cooling followed by limited daytime heating—descends the Rocky Mountains. The air remains cool because the compressional heating fails to sufficiently warm the frigid air before it hits towns along the Front Range. The air may only warm 10 to 15°F on its descent. The topography of the Rockies can channel the airflow, causing it to strengthen to speeds over 60 miles per hour.

From January 30 to February 4, 1989, an arctic air mass invaded the northern Rockies, bringing record cold temperatures and extreme winds to Montana. Ahead of the front, on January 30, downslope winds gusted to 100 miles per hour at Shelby, 102 miles per hour at Cut Bank, 114 miles per hour in Augusta, 117 miles per hour at Browning, and 124 miles per hour at Choteau. The extreme winds toppled twelve empty railroad cars in Shelby, while elsewhere, roofs blew off homes, mobile homes rolled over or broke apart, and numerous trees and power lines blew down. Temperatures plummeted as the front moved through the state: in Great Falls the temperature dropped from 54°F to −23°F and stayed below −20°F for 4 days; in Helena, the temperature remained below −20°F for 3½ days, including a record low of −33°F on the fourth day (windchills dropped the temperature to −75°F); Missoula saw record lows of −22°F and −23°F; and the mercury dropped to −52°F in Wisdom on February 3. The event killed four people. The National Weather Service ranks this arctic outbreak as the fourth most significant weather event during the twentieth century in Montana.

sustained speeds of 30 to 50 miles per hour with occasional lulls in the 20- to 30-mile-per-hour range. Gusts will range from 63 to 84 miles per hour. Although these characteristics relate specifically to Chinooks in Boulder, Chinooks at other locations along the base of the Front Range have similar traits.

Chinooks usually pack the mountains with dark, ominous clouds, but they rarely drop moisture on the Front Range. The abrupt edge of the clouds resembles a wall, so the leeward edge of this orographically induced cloud bank—as seen from the lee side of the mountains—gets the name *foehn wall*. Foehn walls often signal heavy precipitation, especially snow in winter, falling over the Rockies west of the Front Range. They form as a result of orographic lifting. As the air reaches the crest of the Front Range and begins its descent, condensation ceases and the clouds end.

Sometimes lenticular (lens-shaped) clouds form. As swiftly moving air hits the mountains and overtops them, it flows in a wavelike pattern. Like water flowing over a rock in a stream, the wind creates ripples. Lenticular clouds form at the top of the wave crests, where the air ascends just enough to cause condensation. Sometimes a series of lenticular clouds will occur downwind of the Rocky Mountains. In some cases up to four wave crests will form lenticular clouds over Front Range locations. Eventually the waves dissipate and cease to produce more clouds. Individual lenticular clouds often resemble a stack of dinner plates or flying saucers. When a broad, uniform lenticular cloud arches over a large area of the Front Range, it becomes a "Chinook arch." These often remain stationary throughout a Chinook event.

Kelvin-Helmholtz clouds also occur during Chinook windstorms, forming in rows of small curving clouds that resemble cresting waves. The "waves" form as wind just above a cloud moves faster than the cloud. Known as *speed shear*, these winds travel the same direction and have different speeds at different altitudes.

Boulder holds the Colorado wind record. The National Center for Atmospheric Research recorded a Chinook that blew at 147 miles per hour on January 25, 1971. Higher winds have been reported on Longs Peak—about 28 miles northwest of Boulder—and Mount Evans west of Denver, but officials have never accepted these readings of nearly 200 miles per hour. The city of Colorado Springs recently installed a weather station on Pikes Peak, and the University of Denver now has a remote weather station on the top of Mount Evans. Researchers now have the opportunity to examine mountain-peak wind gusts that may someday eclipse Boulder's record.

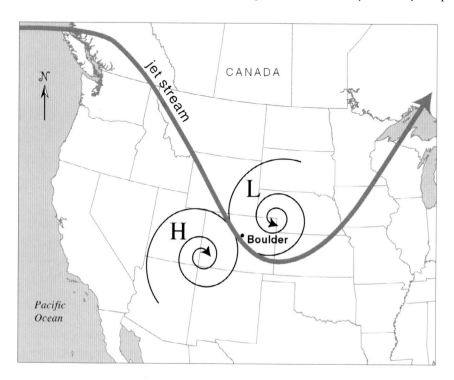

Weather pattern for strong Chinook winds in the Boulder, Colorado, area. Flows around a low-pressure system over western Nebraska and a high-pressure system over the Four Corners meet over Boulder as an onslaught of northwesterly winds. Meanwhile, the northwest-southeast oriented jet stream, which separates the pressure systems and their differing air masses, strengthens the wind. The combined pressure systems and jet stream create a strong northwest wind throughout the entire column of air above Boulder.

Side-view of the dynamics associated with a Chinook in Boulder, Colorado. As winds descend the Rocky Mountains, air compresses and undergoes compressional heating. This causes lower pressure in the lee of the mountains, creating a tight vertical pressure gradient, as seen by the packed lines of equal pressure. These lines can be thought of as streamlines of wind. This causes the winds to accelerate more. At the wave crests, lenticular clouds form as air rises just enough to condense into a localized cloud before heading down the wave and drying out again.

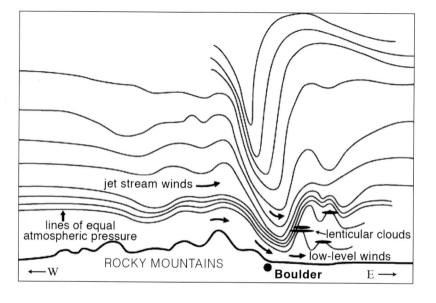

Quickly changing weather patterns and local topography cause strong, often localized, highly changeable, and inconsistent winds along the Front Range. This includes Boulder's winds. The strongest Chinooks occur most frequently between late November and late January, because at this time of year the jet stream tends to travel across Colorado.

During a single year, a 68 percent chance exists that winds near Boulder will gust to or above 90 miles per hour, with a 29 percent chance that a vicious Chinook of 120 miles per hour or faster will jostle the city. Although seasonal, the winds create "flagged" trees across the foothills of the Rockies. Combed by the wind, flagged trees resemble flags

Notorious Wind Events along the Front Range

January 7–8, 1969: Frequent gusts of 100 miles per hour and higher blasted western Boulder between 6:00 and 9:00 PM. Downtown Boulder clocked a gust at 96 miles per hour. Winds caused heavy damage and blew roofs off twenty-five homes.

November 26, 1977: A fierce windstorm blew out windows, downed power lines, ripped roofs from homes, and injured several people. Wind gusts clocked at 109 miles per hour hit western Boulder. A week later Boulder took another battering with winds blowing up to 103 miles per hour. Winds of 90 miles per hour and higher hit Boulder seven times that year.

January 16–17, 1982: One of Boulder's most destructive windstorms of the eighties occurred. Winds gusted up to 120 miles per hour and faster twenty times during the two-day storm. The National Center for Atmospheric Research (NCAR) recorded the highest gust at 137 miles per hour. Forty percent of the buildings in Boulder sustained damage.

January 28–29, 1987: One of the most destructive windstorms of the twentieth century inflicted $5.6 million in damage to the Front Range of Colorado. While Lakewood and Golden received the hardest hit, on January 29 NCAR in Boulder clocked the storm's strongest gust of 99 miles per hour at 8:00 AM. The storm's fury shattered windows, twisted

traffic signals, downed power lines, damaged roofs and cars, and toppled trees and chimneys. Several people sustained minor injuries from flying debris.

January 9, 1989: An early-morning (4:00 to 11:00 AM) Chinook battered the city of Boulder with wind gusts up to 115 miles per hour. Strong winds blew out windows, ripped shingles and roofs off homes, businesses, and schools, downed trees, and flipped several airplanes at the Jefferson County and Boulder Municipal Airports.

January 24, 1992: A 143-mile-per-hour wind gust rocked NCAR, the second highest wind gust ever recorded there. Winds throughout Boulder gusted between 60 and 100 miles per hour, overturning several tractor-trailers, tearing down traffic signals, and damaging roofs.

September 24, 1996: An unusual early-season downslope windstorm ravaged the Boulder area. Winds peaked at 131 miles per hour at NCAR, setting the all-time highest wind speed record during the month of September in Boulder. Gusts of 80 to 100 miles per hour blasted other parts of town. Widespread and severe damage resulted, leaving downed trees, blown-out windows, de-shingled roofs, and cars blasted with sand. At least one person sustained serious injury. Damage estimates soared over $500,000.

Kelvin-Helmholtz clouds above Colorado Springs on Thanksgiving Day, 1999

Standing lenticular clouds above Fort Collins, Colorado, on March 20, 2002

because their branches only grow on the lee side of the prevailing westerly winds.

Though difficult for many of us to imagine, a windstorm can shatter automobile windows and dull paint with blowing sand, and this commonly occurs during a Boulder windstorm. City planners have made efforts to make the city windproof, such as anchoring utility poles. Because of efforts like this, Boulder only sustains serious damage once wind gusts reach 90 miles per hour when even a 60-mile-per-hour storm would devastate most U.S. cities.

Except for the tornadic winds of the Great Plains, Boulder regularly endures the strongest winds of any U.S. city. Even still, Boulder's winds fall short of world records. The highest remotely measured wind speed, taken using a portable Doppler radar, registered 287 miles per hour in an Oklahoma tornado. From the damage a 1957 Dallas twister caused, meteorologists inferred the winds blew at 310 miles per hour—the strongest wind ever witnessed. The mightiest wind gust ever actually measured on Earth blasted at 231 miles per hour atop Mount Washington, New Hampshire, on April, 12, 1934. A wind gust of 236 miles per hour—caused by super Typhoon Paka—was recorded on the island of Guam on December 16, 1997, but after extensive review, officials considered the report unreliable because damage assessment following the storm did not support this wind speed measurement.

Front Range Flash Floods

Flash floods brought about by stalled or slow-moving thunderstorms tend to hit the canyons and foothills along the Rocky Mountain Front from Colorado north to Montana. Moist air rotated north and then west by low-pressure systems

Foehn wall over the Rocky Mountains seen from Loveland, Colorado, during a Chinook event in February 1987

over the Great Plains feeds these flood-producing thunderstorms. The storms pile up, slow down, and even stall along the north-south-oriented Rocky Mountain Front. With a continuous supply of low-level moisture, the storms can last and redevelop over the same location for hours, dropping heavy flooding rains. Furthermore, the higher terrain of the Rocky Mountain Front and the mountains behind them produce orographic precipitation, adding more moisture to the already moisture-heavy storms.

In Colorado, most flash floods occur in July and August when monsoonal moisture streams into the state from the Gulf of Mexico and the tropical Pacific Ocean (see *Southwest Monsoon* in *chapter 6*). At this time of year the upper-level winds of the jet stream weaken to their lowest point, since the temperature contrast between the north pole and the equator decreases. With plenty of available moisture, this region's thunderstorms, which normally occur at this time of year, can bloat and stall in the atmosphere.

On July 31, 1976, the eve of Colorado's Centennial, 2,500 to 3,500 campers and visitors converged in Big Thompson Canyon west of Loveland. The weather forecast had called for the usual summertime weather of scattered afternoon and evening thunderstorms, but otherwise pleasant weather. By 6:00 PM, however, cumulus clouds began to build along the foothills of the Front Range. By 8:30 PM, one of the most destructive and deadly flash floods in U.S. history unfolded.

From 8:30 PM to 12:30 AM, a massive thunderstorm stalled over Big Thompson Canyon—a 20-mile-long canyon between Loveland and Estes Park. Light easterly winds associated with a low-pressure area over southeastern Colorado pushed vast amounts of humid air up the canyon, building, strengthening, and feeding the thunderstorm. Weak, upper-level winds of less than 20 miles per hour moved the storm slowly, allowing it to drop tremendous amounts of rain—10 to 12 inches in parts of Big Thompson Canyon. The

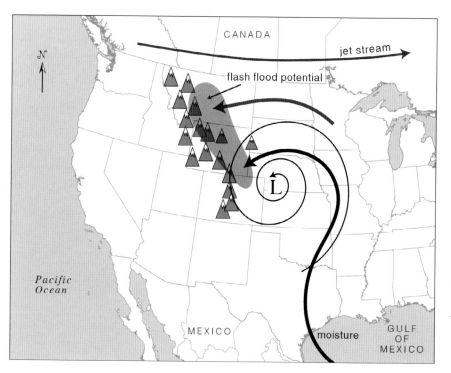

Summertime weather pattern that can cause flash floods along the Rocky Mountain Front of northern Colorado, Wyoming, and Montana. Counterclockwise winds around a low over the central Great Plains pump moist air from the Gulf of Mexico into the region. Light winds aloft and the jet stream well north into Canada allow moisture to remain in the region. Daytime heating creates rising air, which, when supplied with ample moisture, cools, condenses, and turns into flash flood–producing storms.

A house rests precariously on a bridge along U.S. 34 near Drake, Colorado, after rising water carried it down the Big Thompson River. —Photo courtesy of Fort Collins Public Library

One of many sections of U.S. 34 in Big Thompson Canyon washed away in the Big Thompson Flood. —Photo courtesy of Fort Collins Public Library

Montana Flash Floods

On June 22, 1938, torrential rains from several clusters of strong, slow-moving thunderstorms tracked east across Montana, causing flash flooding on the Yellowstone, Musselshell, and Sun Rivers. The worst flooding occurred in the watershed draining the Bears Paw Mountains near Havre. More than 5 inches of rain fell in the Gravel Coulee watershed in one hour, and floodwaters rushed out of the foothills and traveled north 10 miles, eroding 2 miles of Great Northern Railroad track near Laredo. Farther northeast in Havre, floodwaters caused nearly $500,000 in damage ($6.4 million in 2002 dollars) as the normally dry Bullhook Creek became a half-mile-wide torrent of water. Thirty-five miles east of Havre in Zurich, 2 inches of rain fell in one hour and 3.55 inches of rain fell during the entire storm. At least eight people drowned in Zurich, while in Fort Benton, southwest of Havre, twenty-seven small bridges were washed out. The National Weather Service ranks this flash flood as the third most significant weather event during the twentieth century in Montana.

Three days earlier on June 19, a different rainstorm over the Prairie County highlands, north and west of Terry, Montana, produced a flash flood that roared down Custer Creek. The flood weakened a railroad trestle just before an eleven-car passenger train called the *Olympian* crossed. The trestle gave way as the train crossed, and the train hit the far bank of Custer Creek moving at full speed. Of the 140 passengers and crew, 49 perished and 65 were injured. The Yellowstone River carried some of the bodies 130 miles downstream to Sydney, Montana. The National Weather Service ranks this flood as the second most significant weather event in Montana during the twentieth century.

National Weather Service estimated that roughly 8 inches fell in less than two hours near the Big Thompson River's headwaters. The soil and steep canyon walls could not absorb the rain, and the river quickly rose over its banks. Before long the normally 18-inch-deep river became a deadly torrent 19 feet high, crashing through the narrow canyon at 25 feet per second (17 miles per hour). The debris-filled river picked up everything in sight, including 10-foot-diameter boulders, homes, cars, buildings, and much of U.S. 34 at the mouth of the canyon. At 9:00 PM the river flowed at 31,200 cubic feet per second, the power of its 500-year flood (a flood that would statistically have a 0.2 percent chance of occurring during any given year). Just hours before the onslaught, the river flowed at only 137 cubic feet per second. Survivors of the flood described the roar of the river as having sounded like jet engines and called it "the roar of death."

The flood claimed 145 lives, including 6 people whose bodies remain lost. It obliterated 418 homes in the canyon, damaged 138 more, and swept away 52 businesses. The floodwaters dispersed rapidly after exiting the canyon, limiting damage in Loveland. Days later crews dredged at least 197 cars from the mud and rocks along U.S. 34 in western Loveland. The flood removed more than 300,000 cubic yards of mud, rocks, and vegetation from the canyon—enough material to fill a 120-foot-deep hole the size of a football field. Estimated damage from the flood totaled $35.5 million ($112.3 million in 2002 dollars). This flood went into the history books as Colorado's single worst natural disaster in recent history.

Another disastrous flood hit Fort Collins, about 20 miles northeast of Big Thompson Canyon, on July 28, 1997, just days before the twenty-first anniversary of the Big Thompson Flood. A similar weather situation—a stalled thunderstorm with ample monsoonal moisture available—inundated Fort Collins with torrential rains. That day, before the evening storm arrived,

Fort Collins had already received 2 to 4 inches of rain; with soils already saturated, streams ran high when the storm stalled.

The storm centered itself about 1½ miles southwest of town and dropped 10.40 inches of rain between 5:00 and 10:30 PM. By July 29, the Colorado State University campus had received 6.07 inches, the majority of it having fallen in the early hours of the storm. Some areas of the foothills west of Fort Collins received over 10 inches of rain, nearly 70 percent of the normal yearly rainfall for this region. The rains produced flash floods throughout town. Spring Creek, normally gentle, raged the night of the storm, and its banks received the hardest hit. The floodwaters originated from the crest of the Dakota Hogback, a sharp ridge forming the eastern edge of Horsetooth Reservoir in western Fort Collins. The rainfall and runoff from this area—most of the Spring Creek watershed—accumulated and flowed through the middle of Fort Collins. Floodwater overturned cars, swept away mobile homes, flooded base-

ments, tore out bridges, ruptured gas lines, and even derailed a train. Five women died and over thirty-five people received injuries. Damage estimates soared to over $100 million.

Besides Montana and Colorado, Wyoming has also experienced devastating flash floods along the Front Range. The afternoon weather on August 1, 1985—scattered afternoon clouds and thunderstorms—seemed ordinary to residents of Cheyenne, Wyoming's capital. This typifies summer weather for extreme southeastern Wyoming, an area of rolling hills east of the Rocky Mountains. Cheyenne sits atop Cheyenne Ridge—a ridge that extends from Cheyenne eastward to the Colorado-Wyoming-Nebraska border—at 6,062 feet of elevation. By 4:00 PM that fateful day, a tremendous thunderhead began to build over Cheyenne, where the storm later stalled.

By 5:00 PM the thunderstorm had soaked Cheyenne with 0.60 inches of rain, and it continued to unload extremely heavy rain and hail. By 6:00 PM an additional 0.77 inches had fallen,

Aftermath of the July 28, 1997, flood in Fort Collins, Colorado, photographed on the morning after the flood near a trailer park along Spring Creek —John Weaver photo, National Oceanic and Atmospheric Administration

Montana's Top Weather Event of the Twentieth Century

Between June 6 and 7, 1964, a slow-moving line of thunderstorms brought torrential rains to the Rocky Mountain Front in western Montana. Station's along the eastern slopes of the entire Continental Divide in Montana reported record twenty-four-hour rainfall amounts of 8 to 14 inches. The watershed that feeds Gibson Reservoir near the Bob Marshall Wilderness received enough water to fill the reservoir 2 1/2 times. It can hold 99,057 acre-feet of water. The excess water from the reservoir poured into the Sun River, which runs southeast toward Great Falls, where it empties into the Missouri River. Typically the Missouri runs higher than the Sun, so its waters cause the Sun to back up. On June 9 and 10, well after the rains had stopped, however, the Sun, infused with floodwater from Gibson Reservoir, caused the Missouri to back up. The Meadowlark Golf Course and Country Club in Great Falls flooded; nearly three thousand people evacuated as 10 to 12 feet of water spread over the confluence area. Advance warning helped prevent loss of life.

Further north it rained longer, through Sunday, June 7. On the east side of Glacier National Park, Divide Creek spilled over its banks. Logjams along Divide Creek built up and broke, releasing periodic flash floods that knocked out power and telephone lines and stranded tourists in campgrounds and lodges. Two dams failed—Swift Dam on Birch Creek, which held back 30,000 acre-feet of water, and Lower Two Medicine Dam on Two Medicine Creek, which held back 16,600 acre-feet—drowning more than thirty people on the Blackfeet Reservation. Floodwaters breached several irrigation dams on tributaries of the Marias River east of Glacier National Park. Along Birch Creek south of Cut Bank, a 20-foot-high wave devastated Birch Creek valley, destroying all buildings and bridges in its path. Nineteen people died.

Overall, flooding from the two-day rainstorm killed fifty-eight people. It caused one of Montana's most deadly weather-related events on record. The National Weather Service ranks it as Montana's top weather event of the twentieth century.

and then a phenomenal 3.51-inch deluge came between 7:00 and 8:00 PM. By 8:30 another 1.18 inches had fallen, and by 9:00 the National Weather Service office in Cheyenne had measured 6.06 inches of precipitation—an all-time twenty-four-hour record for Wyoming in a little over three hours. Seven inches of rain (measured by a different weather station) flooded the downtown area. This amounts to half of Cheyenne's average yearly precipitation.

Besides rain, hail up to 2 inches in diameter hammered the city. Seventy-mile-per-hour winds flung hail through Cheyenne, defoliating trees, stripping paint from homes and cars, and smashing windows. Floodwaters and winds piled hail into 8-foot drifts in the streets and up to 5 feet deep in basements. Unprecedented runoff caused area streams and rivers to flash flood. The worst flash flooding occurred between 7:00 and 8:00 PM. During the heaviest rains, Dry Creek in northwestern Cheyenne had enough power to float cars and trucks. Ten people, including a sheriff trying to rescue a child, drowned as Dry Creek carried them away in their cars.

As if things weren't bad enough, the National Weather Service issued a tornado warning at the height of the storm when two funnel clouds appeared. Upon hearing the warning, one elderly woman took shelter in her basement only to drown there. Luckily for Cheyenne residents, the funnel clouds lost momentum; they did not touch the ground and did not turn into tornadoes.

In all, the storm killed twelve people in Cheyenne, injured at least seventy, and inflicted $65 million ($108.7 million in 2002 dollars) in property damage. While the storm pounded Cheyenne, residents as far away as Fort Collins, Colorado, watched the enormous thunderstorm light up the night sky. Climatologists and Wyoming residents alike recognize this as Wyoming's most damaging and deadly flash flood in the state's history.

Aftermath of the August 1, 1985, Cheyenne storm, showing hail washed into deep piles by floodwaters —Michael Mee photo, Federal Emergency Management Agency

Several of this country's most deadly and destructive flash floods have occurred along the Rocky Mountain Front. To reduce the deaths, injuries, and damage these floods cause, several Front Range communities, including Denver and Fort Collins, have developed a dense network of rain gauges. These gauges continuously monitor rainfall and have helped forecasters predict floods and warn residents.

Temperature Roller Coaster: Montana

West-central Montana endures some of the world's greatest daily and yearly temperature swings. A surge of extremely cold arctic air, which meteorologists refer to as a Siberian Express, can keep temperatures low, only to be replaced by warm Chinook winds sweeping down the eastern slopes of the Rocky Mountains. The reverse can happen as well: arctic air can replace Chinooks.

In winter when high pressure resides over the West Coast and low pressure over the Ohio Valley, west-central Montana experiences Chinooks (see *Snow Eater* earlier in this chapter). This pattern can exist for several days, providing Montana with warm, windy weather. Eventually, a strong storm will move across the Pacific Ocean,

and the high-pressure system deflects the storm northeastward. As the storm moves into central British Columbia, its circulation gathers arctic air from the cold Canadian prairies and it regains its strength after having been weakened by the high. Following the high's clockwise circulation, the storm dives south through British Columbia along the eastern side of the high. This causes the high-pressure system, relatively weaker at this point, to retrograde—move in the opposite direction of the basic flow in which it is embedded. These events "open the door," as meteorologists say, for the storm to move into the United States. Retrograding weather systems, for both high- and low-pressure systems, are not unusual, and they generally occur with large storm systems.

When the high retrogrades, the warm northwesterly Chinooks over Montana shift to a northerly direction, and arctic air plunges southward into the state. As the dense arctic air undercuts the warm Chinook layer at lower elevations, the temperature drops at the rate of several degrees Fahrenheit per minute. An inversion occurs—cold air under warmer air. Meteorologists call it a *frontal inversion* since it borders two different air masses (a front) and is caused by an inversion. The Chinook

Jim Wood, a Cooperative National Weather Service observer, near his cotton region weather shelter in Loma, Montana. This shelter registered the nation's greatest twenty-four-hour temperature change of 103°F between January 14 and 15, 1972.
—Photo courtesy of the National Weather Service, Great Falls, Montana, Office

continues to blow above the arctic air but loses strength due to the shift in wind direction.

Likewise, surface temperatures can change dramatically when Chinook-warmed air suddenly replaces an arctic air mass. This occurs when a dense and often shallow layer (only a few thousand feet deep) of frigid air hovers over the Rocky Mountain Front. High pressure rebuilds across the West as a cold air–producing storm moves east across the Great Plains and upper Midwest. As this happens the winds aloft slowly shift from the cold northerly direction back to a warmer northwesterly or westerly direction, bringing a relatively warm, downslope Chinook off the Rockies.

Initially the Chinook flows over the cold surface layer. The compressionally warmed air caps the cold air mass, creating a frontal inversion. The inversion creates a stable atmosphere and holds the cold air in place, preventing it from mixing with the warm air above it, so the winds in the cold air mass remain light and variable. As the low continues tracking east, the Chinook winds strengthen and break down the frontal inversion, allowing warm air to rush to the surface. This can lead to remarkable warming, sometimes on the order of several degrees Fahrenheit per hour.

Often a battleground for dramatically differing air masses, west-central Montana has become famous for some of the world's most spectacular daily and yearly temperature differences. Loma, Montana, downwind of the Continental Divide approximately 50 miles northeast of Great Falls, holds the nation's record for greatest temperature change in twenty-four hours. The temperature at 9:00 AM January 14, 1972, reached –54°F, but then conditions changed as a Chinook developed; by 8:00 AM the next morning, the temperature climbed to 49°F! An astounding difference of 103°F! This broke the old world record set at Browning, Montana, just east of Glacier National Park on the Blackfeet Reservation. On January 23, 1916, a warm Chinook wind boosted the temperature at Browning to a balmy 44°F, but the next day a vigorous arctic front dropped the temperature to –56°F—a drop in temperature of 100°F in less than twenty-four hours.

Fairfield, Montana, 30 miles northwest of Great Falls, cooled from an unseasonably warm 63°F to a bone-chilling –21°F in just twelve hours on December 24, 1924—a 7°F drop per hour. This stands as the most dramatic twelve-hour temperature drop ever recorded in the world.

An almost identical event occurred in Granville, North Dakota, in 1918, but it missed the world record by 1°F.

Montana's Temperature-Rise Record

On January 11, 1980, the temperature at Great Falls International Airport rose from −32°F to 15°F in seven minutes as a Chinook eroded an arctic air mass. This 47°F rise stands as the most rapid temperature change ever recorded in Montana.

Glasgow, Montana, holds the U.S. record for having experienced the greatest temperature range in a year. In 1936, Glasgow's temperature extremes ranged 171°F: −59°F on February 15, and 112°F on July 18. The all-time extreme daily temperature range across the West is an astounding 204°F: from 134°F in Death Valley, California, to −70°F at Rogers Pass, Montana.

Remarkably, the temperature in Montana has ranged from a hostile −70°F recorded at Rogers Pass (see *Coldest Place in the Lower Forty-Eight* in *chapter 8*) to a blistering 117°F. This phenomenal 187°F difference stands as the U.S. record for the greatest difference between the highest and lowest temperatures ever recorded in a state. Not surprisingly, Montana is often called the "land of extremes." Only Utah and North Dakota have felt a similar span of temperatures. As for the entire planet, the all-time maximum and minimum temperature extremes span 265°F: −129°F and 136°F recorded at Vostock, Antarctica, and Al-'Aziziyah, Libya, respectively.

Back Door Cold Fronts

Cold fronts in the United States normally move from north to south or west to east as they follow the polar jet stream, always encircling the earth in a westerly direction. Occasionally, however, a cold front will drift backward. Meteorologists call these backward-drifting cold fronts, which are common around the world, *back door cold fronts.*

Meteorologist most often use this term when a high-pressure system over New England moves cool Atlantic air into the northeastern United States, replacing warmer continental air. Along the Rocky Mountain Front, however, winter back door cold fronts separate air masses with vastly different temperatures, making these events seem particularly remarkable. Although back door cold fronts can occur year-round along the Rocky Mountain Front, they occur most often in winter, when cold fronts drop down from Canada with regularity.

Mountainous Snowdrifts

On April 25–26, 1969, a late-season storm brought a drastic change in weather to eastern Montana. A day after numerous stations registered their highest temperature of the month (many in the 80s) as a result of a Chinook, an arctic cold front swept through Montana, bringing blizzard conditions to much of the eastern half of the state. Temperatures fell more than 50°F in twenty-four hours with windchill readings dropping much of eastern Montana below zero for nearly forty-eight hours. Snowfall totals for much of the region measured over 1 foot, with higher totals at certain locales, including 32 inches near Sonnette. Wind whipped the fresh snow into 20-foot drifts in places. Power and phone lines went out over a twelve-county area, with some southeastern residents going without power for two weeks and without telephone service for over a month! The storm inflicted nearly $2 million ($9.8 million in 2002 dollars) in property losses, excluding animal losses. Over 100,000 sheep, horses, and cattle died—in today's dollars totaling well over $10 million. The National Weather Service ranked this severe storm as the seventh most significant weather event of the twentieth century in Montana.

Back door cold fronts develop in the wake of a low-pressure system and its arctic cold front. A cold front drops down from Canada behind a low, passing east of the Rocky Mountain Front and sparing the region frigid temperatures. As the front sweeps southeastward across the Great Plains and the Midwest, a cold high-pressure system develops over the Dakotas in response to the cold air. As the low and its front continue east, the high grows stronger and larger. Eventually the clockwise rotation of the high pushes an elongated cold front westward—a back door cold front— across Wyoming, Colorado, and Montana and up against the foothills of the Rocky Mountains, causing a sudden drop in temperatures.

As the back door cold front reaches the Rocky Mountain Front, it stalls and can move no farther. The shallow (a few thousand feet deep)

and dense cold air behind (east of) the front cannot move over the higher terrain. This leaves the higher elevations in the Rocky Mountains west of the cold front much warmer than the eastern lowlands. The temperature contrast across the front can rise to 50°F. Back door cold fronts tend to be dry since they originate from dry northern locales; however, clouds and snow flurries can occur along and just east of the cold front as enough air is forced high enough up the terrain to induce condensation. As the high-pressure system continues moving south, it pushes the elongated back door cold front, which continues getting longer, southwestward, in almost the opposite direction a cold front usually travels in this region.

Because the Rocky Mountains are oriented from the south to the northwest across the West, southwesterly-moving back door cold fronts back

a.

b.

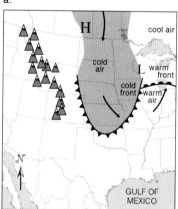

c.

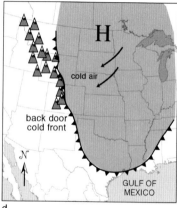

d.

Development of a back door cold front. (a) A low-pressure area moves into the northern United States, with a cold front trailing behind it ushering much colder air into the northern Great Plains. (b) The storm continues its southeastward movement, sparing the Rocky Mountain Front from cold temperatures as the cold front pushes southeastward. (c) High pressure begins to build behind the low-pressure system, now moving into the Great Lakes region and still sparing the Rocky Mountain Front from any cold air. (d) The high-pressure system moves into the northern Great Plains, and its clockwise flow forces a cold front—a back door cold front—up against the Rocky Mountain Front of Colorado, Wyoming, and southern Montana.

up against them and stall. Some of the mountains in Wyoming, particularly the Bighorns, occur farther east than the Rockies in Colorado or Montana, and therefore lie in the path of more back door cold fronts than other places along the Rocky Mountain Front. When these fronts linger on the east side of the Bighorns, they bring snow flurries and fog to the region. At the same time, to the west, relatively warm above-freezing temperatures and scattered clouds prevail.

A typical back door episode occurred on the afternoon of November 22, 1996. A cold front swept south out of Canada and then west up against the Bighorn Mountains and the Front Range in north-central Wyoming. Consequently, Casper, Wyoming, west of the Front Range, reported a relatively balmy high temperature of 59°F, while just 109 miles to the north and across

the back door cold front, Buffalo, Wyoming, only managed to reach 21°F—a temperature difference of 38°F.

Back door cold fronts don't always make it to the Rocky Mountain Front. If the northeasterly winds from the Great Plains high pressure lack strength, back door cold fronts sometimes stall out over eastern Wyoming and Colorado. They often dissolve or revert to typical eastward-moving cold fronts after a day or two, but back door cold fronts can hover for much longer periods, up to a week in extreme cases, keeping temperatures low. When higher pressure moves east into the Rocky Mountain region, the cold front begins moving east again.

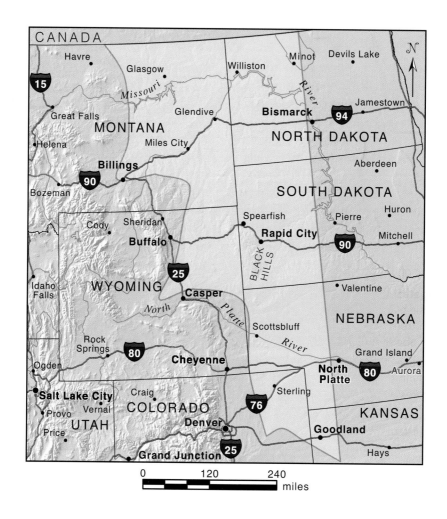

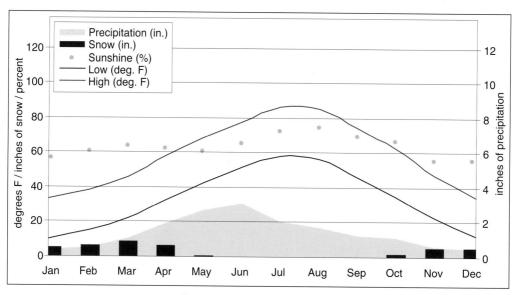

Climate of Rapid City, South Dakota

13

Northern Great Plains

Elevation: 1,500 to 7,000 feet

Natural vegetation: grass and other herbaceous plants; scattered needleleaf evergreen trees in the higher elevations (generally above 5,000 feet)

Köppen climate classification: dry, particularly in the summer, with cool temperatures

Most dominant climate feature: extremely cold winter temperatures

Besides the cold western valleys in the Rocky Mountains, the northern Great Plains and its islandlike mountain ranges composes one of the coldest regions of the country during winter. Although portions of the eastern Dakotas may experience colder temperatures, the western portion of North and South Dakota along with northeastern Montana share similar brutally cold winters. In January and February, daytime temperatures usually warm into the 20s in the north (Montana and North Dakota), and into the 30s in the south (western Nebraska and western Kansas), with overnight lows from −5 to 15°F. Temperatures can range wildly and suddenly during a single day—up to 50°F. Other than a few snowstorms delivering light snow, precipitation during January and February scarcely occurs, generally totaling less than ½ inch. The Black Hills, however, usually pick up 1 inch of precipitation (including melted snow).

The winter thaw begins in March and April. Although it can still get cold in March with daytime readings in the 30s and 40s, daytime highs reach the 50s and 60s in April. Overnight lows through the period range from the 20s in March to the 30s in April. Winter storms are common but often track to the south, leaving much of the region with little snow. Total precipitation for the March-April period ranges from ½ inch in the north to locally 3 inches in the south.

The last spring freeze often occurs in May. Considerably warmer, May and June usher in warming daytime temperatures; they can soar into the 90s by June. Daytime readings in the 70s are more typical, though, and overnight lows dip into the 40s and 50s. Quite wet, May and June constitute the wettest months of the year, averaging 2 to 7 inches of precipitation, which falls heaviest in the mountains. Although it can still snow, especially in the north and across the higher terrain (generally above 5,000 feet), most moisture falls from thunderstorms or showers associated with cold fronts sweeping down from the north or in from the west. A few times each year thunderstorms produce large hail, strong winds, and flooding rains in June. Although rare, tornadoes do occur at lower elevations (generally less than 3,000 feet).

July and August bring similar weather conditions; daytime temperatures reach to the 80s and low 90s, overnight lows drop to the 50s, and average precipitation stays in the ½- to 2-inch range. Precipitation falls from thunderstorms initiated by cold fronts and intense surface heating.

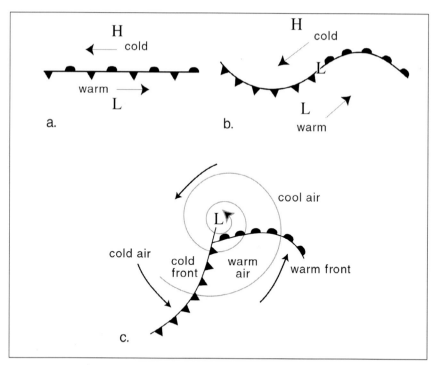

a.

b.

c.

The Norwegian Cyclone Model, develop in the early twentieth century, describe general formation of mid-latitude storm systems. Its creators hoped it would help improve weather prediction and forecas. The model includes three stages: initiati development, and the mature phase. (a, Initiation: A stationary front separates differing air masses—cold to the north and warm to the south. The clockwise ; around the high pressure of the cold air mass causes its winds to blow from east west. Lower pressure and counterclockw winds in the warmer air mass cause winds to blow from west to east. (b) Development: Differing wind direction pressure cause a disturbance, or eddy, to form along the stationary front, deform. it into a distinct warm front and cold front. A strong low-pressure system form. where they intersect. (c) Mature phase: the low intensifies, a well-organized sto. system develops.

The biggest weather threat to property and crops during this two-month period comes with hail dropped from the strongest storms; large hailstones, up to 2 inches in diameter, can occur.

By mid-October the first fall freeze occurs and temperatures begin to drop. Although September can have mild daytime highs in the 70s and low 80s, October quickly cools into the 60s during the day and 30s and 40s at night. September has slightly more precipitation than October, but as a whole each month stays relatively dry with a two-month precipitation total ranging from 1 to 2 inches. The mountains receive the majority of the precipitation.

Cold and windy, November and December commonly bring light snow. Storms usually don't drop more than a few inches. Although heavy snowfall amounts rarely occur, blizzards can drop 3 inches of blinding snow, halting all transportation. The snow that does fall during November and December often stays on the ground throughout the winter. In November daytime temperatures remain in the 40s but drop

to the 20s in the north and the 30s in the south during December. The light snow that falls typically equates to less than 1 inch of precipitation for the two-month period.

Deep Freezer

Its intercontinental location makes the northern Great Plains very dry and susceptible to extreme, long-lasting low temperatures. The dry and sparsely vegetated plains lie in the rain shadow of the Rocky Mountains, so the Pacific storms that do pass through this region don't usually drop much precipitation; wetter weather is usually associated with storms passing across Wyoming that draw moisture from the Gulf of Mexico. Influenced by passing storms and the jet stream, arctic air masses envelop this region regularly.

Every winter, particularly during La Niña years, many outbreaks of arctic air strike the northern Great Plains. Without any significant east-west-oriented mountain ranges between eastern Montana and the Northwest Territories of Canada, arctic air plunges southward and settles over the northern Great Plains. During the long northern nights of winter, the subarctic atmosphere radiates heat to outer space. The average temperature grows colder each night during an outbreak. As the air cools, it grows denser and heavier, causing an atmospheric high-pressure area to develop as cold air sinks to the earth's surface. A boundary, called a *stationary front,* develops between the mass of cold northern air and milder air to the south. Since the cold air has characteristics of a high-pressure system, its winds blow from east to west. The opposite holds true for the rising air of the low-pressure area; the winds blow west to east. An area of increased wind and wind shear between these two air masses, known as a *jet stream maximum,* causes a disturbance, or eddy, to form along the boundary. This eddy turns the stationary frontal boundary into a cold front, which moves southeastward; a warm front, which moves northeastward; and a strong low-pressure area where they meet—a typical mid-latitude storm system.

As the low moves south and east, its counterclockwise circulation pulls a pool of cold air in behind it. The low pushes warmer air east and north ahead of it. Cold air rushes south behind the cold front, which is characterized by a narrow band of gale-force winds, driving snow, and rapidly falling temperatures. Sometimes blizzards form at the front where the relatively warm air mixes with the arctic air.

Montana's Record Chills

Fort Keogh, Montana, at one time the largest army outpost in Montana, held the lowest temperature record for the contiguous United States before the −70°F reading at Rogers Pass in Montana surpassed it (see *Coldest Place in the Lower Forty-Eight* in *chapter 8*). Located 2 miles south of Miles City, between 1877 and 1908 Fort Keogh served as a major post at the close of the Indian Wars, and on the morning of January 13, 1888, a soldier measured a temperature of −65°F.

Eventually, as a result of this storm system, cold air envelops a large part of the United States. Its coverage largely depends on the extent of snow cover, or lack thereof, over which the air mass travels. Snow cover prevents the sun from warming the earth's surface, which would heat the air mass; therefore the cold air maintains its frigidness, sometimes for long stretches of time. During the dead of winter (January and February), North and South Dakota regularly shiver with daytime temperatures in the 20s and morning lows near zero, while areas south of the cold front enjoy temperatures 10°F warmer.

This area stays cold for long periods of time because a steady supply of storm systems pull in arctic air and also because the polar jet stream, which separates colder air to the north and warmer air to the south, frequents the area north of this region in winter. Even after a cold front has passed, temperatures remain cold as high pressure—induced by the cold air—rotates arctic air in from the north. Warming occurs when a high-pressure system drifts southeast of the region, drawing relatively warmer air in from the south. The strongly bimodal temperatures in the northern plains generally range much warmer or much colder than the average. With most arctic outbreaks, high temperatures struggle to get out of the single digits, while midwinter highs in the 20s can feel balmy.

Temperature Flip-Flops: South Dakota

Known for dramatic temperature changes in very short periods of time, western South Dakota's unique mixture of topography and weather has produced some of the earth's most remarkable temperature fluctuations. The only major mountain range in South Dakota, the isolated Black Hills, sits in the direct path of arctic air outbreaks. The relatively high elevations of the Black Hills, however, are spared from the frigid temperatures of these low-lying air masses, while the remainder of the northern Great Plains suffers.

A cold morning in the Great Plains during the February 4, 1989, arctic outbreak

The Cold Wave of 1936–37

No single cold air event compares to the length and harshness of the winter of 1936–37 in the northern Great Plains. The entire state of Montana suffered extreme conditions—22.4°F colder than normal—and recorded extremely low temperatures: –53°F at Summit, –57°F at Cascade, and –59°F at Frazer and Glasgow. Fifteen people died of hypothermia, and the cold snap set several northern Great Plains records.

All-time coldest average monthly temperature for the lower forty-eight: –19.4°F measured in February at Turtle Lake, McLean County, North Dakota

All-time coldest average winter temperature in the lower forty-eight: –8.4°F measured between December 1936 and February 1937 in Langdon, Cavalier County, North Dakota

Longest duration of days with a maximum daily temperature below 0°F in the lower forty-eight: 41 days, measured between January 11 and February 20 in Langdon, North Dakota

All-time lowest temperature ever recorded in South Dakota: –58°F measured February 17 in McIntosh

All-time lowest temperature ever recorded in North Dakota: –60°F measured February 15 in Parshall

Dakota Dry Spots

The midcontinental location of the Dakotas requires storms to reach a long distance—to the Pacific Ocean or the Gulf of Mexico—for moisture. Even when they do reach a source of moisture, the storms usually only tap a limited supply. Since South Dakota sits slightly closer to the Gulf of Mexico, the primary moisture source for the northern Great Plains, it is slightly wetter than North Dakota. The average annual precipitation in South Dakota is 18 inches, while North Dakota receives 16 inches.

The wet season for the Dakotas runs May through August, when energized storms tap moisture from the Gulf of Mexico and carry some of it from the Pacific Ocean. Most of the summer precipitation falls in the form of rain from evening thunderstorms. During June, the wettest month for the Dakotas, it rains an average of eight to sixteen days. Statistically, the greatest chance of rain for this region (60 percent) occurs on June 29.

Winters stay especially dry in the Dakotas as cold, dry air masses seep down from the Arctic deserts. The driest area of South Dakota, near the intersection of Montana, North Dakota, and South Dakota, sees only 12 to 13 inches of precipitation each year. The driest area of North Dakota stretches from near Williston southwest to near Fairview, Montana, where an average 13 to 14 inches of precipitation falls each year.

In winter, shallow arctic air masses driven by the jet stream surge southward from northwestern Canada (see *Deep Freezer* in this chapter). This shallow (generally between 1,000 and 4,000 feet deep), bitterly cold, dense layer of air fills the valleys of eastern Montana and the Dakotas. As it moves south it plows under a relatively warmer air mass and replaces it. Deeper arctic air masses associated with storms tend to push warm air out of the way and completely replace it; however, shallower surges of arctic air force the warm air mass up and over until it covers the arctic mass like a blanket.

Light upper-level winds above the cold air mass and warm downslope Chinooks flowing off the Black Hills further heat the displaced air mass, increasing the temperature difference between the two air masses. The Chinooks can't penetrate the cold air hovering near the ground, and sometimes they cause the pool of cold air to slosh back and forth like water in a bowl throughout the valleys of the Black Hills. The Chinooks push the cold air east, and when they weaken, the cold air moves west again. This sloshing triggers unbelievable temperature swings as the horizontal boundary of the two air masses moves up and down the sides of South Dakota's valleys. The well-defined line delineating the two air masses can meander up and down the slopes of the Black Hills for hours, even days, until the air masses mix sufficiently and the boundary becomes less discernible.

An extreme example of this type of well-defined temperature boundary—perhaps the most extreme example in the world—occurred in western South Dakota. On January 20, 1943, Lead, South Dakota, perched on a steep hill at 5,400 feet, had a temperature of 52°F while just 4 miles downhill (along U.S. 85), the town of Deadwood shivered at −16°F. The elevation difference between the two towns is only 600 feet, but the temperature difference was an astounding 68°F! Like an invisible wall, the well-defined boundary of the two air masses separated the towns.

On January 22 that same year in Spearfish, which rests at 3,643 feet, the temperature rose and dropped so dramatically that plate glass windows cracked. Temperatures at 7:30 AM measured −4°F in Spearfish as a cold air mass engulfed the region; by 7:32 AM, however, thermometers measured a balmy 45°F as the cold air mass sloshed to the east, and the boundary between the cold and warm air masses moved to lower elevations, allowing the warmer air to reenter the region—an

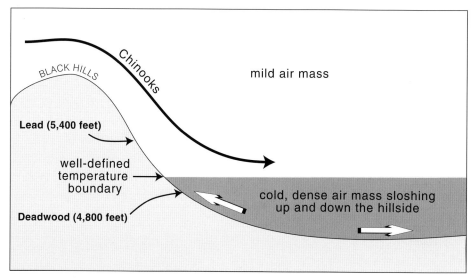

Temperature flip-flops in the Black Hills of South Dakota. Chinooks displace the arctic air mass, causing drastic temperature fluctuations as the well-defined boundary between the warm and cold air masses moves around. Meteorologists refer to this boundary as a frontal inversion since an inversion—cold air under warmer air—causes the front to develop.

astounding 49°F rise in two minutes! By 9:00 AM the temperature had risen to 54°F. For a brief time, Spearfish residents enjoyed a break in the cold weather. It didn't last. The cold air sloshed back up and over Spearfish, sending the temperature to –4°F in twenty-seven minutes. The 49°F rise still ranks as the world's greatest temperature rise in two minutes. Likewise, Rapid City felt amazing temperature gyrations on the same day. Try dressing for weather like that!

Often shallow and heavy, arctic air masses oozing southward into the United States usually measure no more than 4,000 feet thick. Initially they do not mix with the warm air above because of their weight and sinking motion. Eventually the different air masses are mixed by strong, energetic winter winds associated with incoming storms or the eastward movement of the high-pressure system that brought the arctic air in the first place. More often than not, arctic fronts make it all the way to the Gulf of Mexico, but the air mass has warmed significantly by then; meteorologist call this *modified artic air.*

Rapid City Drops

The world's most rapid two-hour temperature drop occurred on January 12, 1911, at Rapid City. The temperature fell from a balmy 49°F at 6 AM to a frigid –13°F by 8 AM, a 62°F temperature drop! The nation's fastest fifteen-minute temperature drop occurred at Rapid City on January 10, 1944, when the temperature plunged from 55°F at 7:00 AM to 8°F at 7:15 AM.

Rapid City One-Hundred-Year Flood

June 9, 1972, started out foggy and humid, but otherwise as a typical summer day in the Black Hills of South Dakota. By late evening, however, the most deadly and controversial flash flood in recent U.S. history roared through Rapid City after a relentless stream of thunderstorms drenched the area.

Rapid City lies at 3,247 feet on the eastern edge of the Black Hills in southwestern South Dakota. In summer, it typically experiences partly cloudy skies, warm days (70s and 80s),

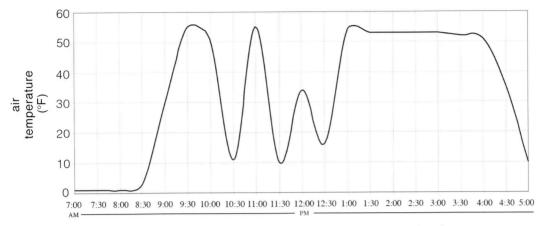

Temperature profile of Rapid City, South Dakota, on January 20, 1943, when the temperature fluctuated drastically as a cold air mass sloshed around the region

cool nights (50s), and scattered thunderstorms. Since upper-level westerly winds lacked strength on June 9, tremendous amounts of humid air from the Gulf of Mexico poured into the area. This massive low-level flow of moisture-laden air, 3,000 feet thick and 78 miles wide, rushed up the eastern side of the Black Hills. These unusually tropical atmospheric conditions, combined with cold air in the upper atmosphere, created an unstable atmosphere. As a result, the headwaters of Rapid Creek, which runs in a southeasterly direction through Rapid City, received a continual barrage of thunderstorms. What meteorologists refer to as *training* occurred over the headwaters: thunderstorms line up like the cars of a train, hitting a region one storm after another, dumping a dangerous amount of rain.

More characteristic of tropical thunderstorms than mid-latitude thunderstorms, these storms had dew points in the upper 60s—common in the tropics but 10 degrees higher than normal for a summer afternoon in Rapid City. This meant a lot more moisture in the atmosphere than normal. Similar to the storm system that caused the Big Thompson Flood in Colorado (see *Front Range Flash Floods* in *chapter 12*), this storm system stalled over the region and dumped

massive amounts of rain. The intense rain of the thunderstorms began at about 6:00 PM. A weather station at Galena, north of Rapid City in Lawrence County, reported 4 inches of rain in two hours.

The Johnstown Flood

The most deadly flash flood in U.S. history occurred on May 31, 1889. Extremely heavy rains coupled with the failure of South Fork Dam sent a massive wall of water through Johnstown, Pennsylvania, killing over 2,200 people.

By 8:00 PM townspeople became uneasy about the weather. Rapid Creek and Box Elder Creek continued to rise, and the rain grew more intense. Utility crews raced to shut off gas lines, people started evacuating Rapid City, and weather forecasters noticed frightening indications of a one-hundred-year flood—rainfall amounts exceeding 10 inches and radars showing slow-moving storms.

Waterlogged thunderstorms continued to develop along a cold front, which extended in an easterly direction along the South Dakota–Nebraska state line. The counterclockwise winds

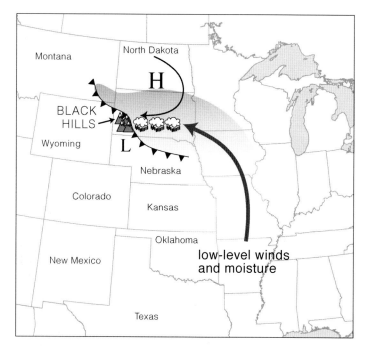

General weather pattern of the June 9, 1972, Rapid City One-Hundred-Year Flood. A cold front draped along the eastern slopes of the Black Hills, mixed with moisture flowing in from the Gulf of Mexico and rising up the Black Hills, caused a long-lasting line of saturated thunderstorms to develop.

associated with a low-pressure area south of the Black Hills swept each individual storm westward into the Black Hills. Time after time, the same areas were drenched by slow-moving thunderstorms. In five hours storms dumped 10 to 15 inches of rain over a 60-square-mile area. Nemo, just north of Rapid City, reported 15 inches of rain during the storm; Rapid City reported 4. Hydrologists estimated that 143 billion gallons of water soaked the Black Hills that night.

Around 10:45 PM, just five minutes after Rapid City officials gave an evacuation order, the earthen Canyon Lake Dam in the western portion of town gave way, adding to the already swollen waters of Rapid Creek. By 11:15 PM, a 5- to 10-foot-high wall of water and debris raced toward Rapid City, which had plunged into darkness since power lines had snapped. Horrified residents rushed for higher ground as the wave of black water smashed through town. Two hundred thirty-seven people died. Hydrologists with the National Weather Service estimated that at the peak of the flood, some 375,000 gallons of water gushed through the city.

Make It Rain

Left to nature, many clouds fail as efficient precipitators and retain more than 90 percent of their moisture. Cloud seeding, however, can greatly improve a cloud's precipitation efficiency. In the United States, cloud-seeding research began in 1946 at the General Electric Research Laboratories in Schenectady, New York, and cloud-seeding projects have taken place in over twenty countries. In the 1950s water resource agencies in the United States began seeding clouds in the attempt to deliver rain to drought-stricken areas. Officials generally conducted projects in winter to increase mountain snowpack or during summer to increase precipitation and/or reduce hail damage. Proponents of cloud seeding believe it can bring relief to dry-land agriculture during times of drought; opponents, however, believe it can lead to deadly floods. The practice remains controversial.

A heap of cars carried downstream by the Rapid City One-Hundred-Year Flood —Photo courtesy of the National Oceanic and Atmospheric Administration/Department of Commerce

To seed a cloud, particles acting as freezing nuclei must be injected into the clouds using an airplane or a ground-based generator. Water vapor adheres to the injected particles and falls to the ground as rain or snow. For seeding to work, clouds must extend high into the very cold atmosphere, where temperatures remain below freezing. The injected particles and moisture must be supercooled—exist in the liquid state at temperatures below freezing—for condensation to occur. At first, cloud seeders dropped crushed dry ice pellets (carbon dioxide) into the top of a cloud from an airplane. Later, they discovered that silver iodide worked better as a cloud-seeding agent. This brought a rainmaking boom in the 1950s, particularly in drought-stricken areas.

Crystals of silver iodide, a salt, serve as good artificial ice-nuclei, attracting moisture from the surrounding air and forming crystals that grow large enough to fall to the ground as snow. Some projects in the West, using ground-based silver iodide generators to seed winter storms over mountain areas, have operated continuously since 1950. Snow-making operations at ski areas work differently. They use machines that break water into small particles, cool it to the freezing point, and distribute it as man-made snow, which is nothing more than tiny ice particles.

Cloud seeding became very controversial when some people blamed it for the Rapid City One-Hundred-Year Flood. Although no one could establish a direct link between the two, no further cloud seeding has been done in the Rapid City area. Despite its limited success, some studies show a slight increase in snowfall in areas up to 90 miles downwind from the ground generators, making cloud seeding well worth the effort to proponents.

Good Lift

Not only are the Black Hills the highest terrain in South Dakota, but also they collect the most precipitation of any place in state. There are three key mechanisms that can force air to rise, cool, condense, and precipitate: convection (air lifted by rising warm air), convergence (air piles up in a region and is forced to rise), and topography. The Black Hills provide lift, technically referred to as *orographic lift,* which results in orographic precipitation.

The hills range in elevation from about 5,000 feet to 7,242 feet atop Harney Peak, the highest point in South Dakota. The average annual precipitation on Harney Peak is nearly 30 inches, over twice the amount the lowlands receive. Rapid City receives an average of 16.10 inches of precipitation a year, 47 percent of which falls from orographically induced showers and thunderstorms between May and July.

Although orographic precipitation occurs in all high mountains, the Black Hills aren't cluttered by other mountains and offer a good example of the ability of mountains to enhance precipitation. Since the moisture for the orographic precipitation normally originates from the south and east in South Dakota, the heaviest snowfall occurs on the south- and east-facing slopes.

By 2:00 AM on June 10, Rapid and Box Elder Creeks began to recede, unveiling a horrible scene of mangled bodies, damaged cars and buildings, and twisted trees. More than 2,800 families lost their homes, and the storm caused $165 million ($711 million in 2002 dollars) in property damage. It became known as the Rapid City One-Hundred-Year Flood, the worst natural disaster in South Dakota history measured by the damage it inflicted and the number of lives it took.

Rapid City residents alleged the flood resulted from experimental cloud seeding conducted by the South Dakota School of Mines and Technology under a contract with the Bureau of Reclamation. Although a direct link was never established between the flood-producing storms and the cloud seeding, no further cloud seeding has been done in that area.

Hail Alley

Although thunderstorms produce dramatic tornadoes and flash floods, hail can do far more damage to property and crops. Hail destroys well over a billion dollars worth of crops each year in the United States. Wheat, cotton, corn, soybeans, and tobacco receive the most damage because these tender plants are often grown in hail-prone areas. Although hailstones have been found that weighed as much as 1.67 pounds, much smaller hail can destroy crops, turning plants to mush in a matter of minutes.

In the United States, hard, damaging hail most commonly falls in the area known as Hail Alley, which includes Nebraska, eastern Colorado, southeastern Wyoming, southern South Dakota, Kansas, Oklahoma, and northern Texas. Hail pelts parts of this region an average of three to ten days a year. It hammers Cheyenne, Wyoming, an average of ten times a year, causing meteorologists to refer to it as the Hail Capital of the United States. Hail Alley, part of a larger area of North America known as the Hail Belt, extends from Hailstone Butte, near Calgary, Alberta, southeastward along the Rockies to northeastern New Mexico and as far east as the Dakotas. This region experiences large hail because of its elevated terrain and high frequency of thunderstorms rolling off the Rocky Mountains. The central Rocky Mountains experience more hailstorms than places in the Hail Belt—perhaps fifteen or twenty a year—but these storms produce smaller and softer hailstones, and often the storms go unnoticed since they cause little if any damage.

When the sun heats the earth during the day, the air close to the ground warms as well. Less dense and lighter than the cold air above it, the hot air rises and cools. As it cools, its capacity for holding moisture decreases, until it can no longer hold all of its moisture. Condensation occurs, creating large, thunderstorm-forming cumulonimbus clouds—the clouds that hail forms in. The condensing moisture releases heat, known as *latent heat,* into the surrounding air, which causes the air to rise faster and shed more moisture, growing into even larger and more powerful storm clouds.

A cumulonimbus cloud contains massive amounts of energy in its updrafts and downdrafts, which can reach speeds in excess of 110 miles per hour. Hail forms in a storm cloud's main updraft, where most of the cloud exists as supercooled water. Without impurities water can remain in its liquid state as it cools below its freezing point; it's referred to as *supercooled water.* Supercooled water will remain a liquid until an appropriate nucleus comes along, such as dust, ice crystals, or salt, to which water droplets will

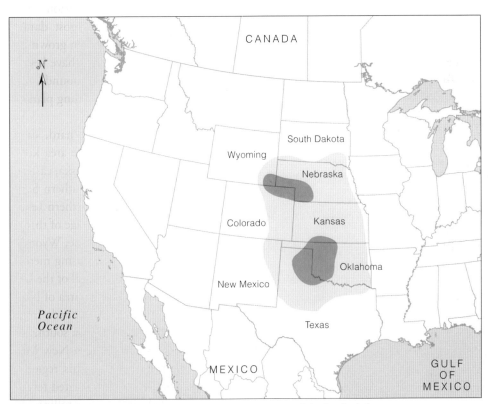

Map of Hail Alley highlighting two areas with higher hail frequency

Number of Severe Hail Events at Cheyenne, Wyoming, 1892–1996*

	MARCH	APRIL	MAY	JUNE	JULY	AUGUST	SEPTEMBER	OCTOBER	NOVEMBER	TOTAL
Total	7	44	214	264	167	122	62	23	1	904
Severe	0	0	19	26	25	12	4	0	0	86

Source: Jack Daseler, National Weather Service, Cheyenne, Wyoming
*Severe hail measures ³/₄ inch or greater in diameter.

Large hail collects on streets and grass during a severe thunderstorm. —Photo courtesy of the NOAA Photo Library, NOAA Central Library; OAR/ERL/National Severe Storms Laboratory

Hail (whitish streaks) falls from a Great Plains thunderstorm as it moves across Washington County in Colorado. —Charles Bustamante photo

freeze. At temperatures above −40°F, supercooled water droplets freeze to these airborne particles. (At temperatures below −40°F, snow forms instead.) While tumbling within the thunderstorm, the frozen droplets continue to collide with more water droplets and grow in size, adding layers like an onion. The once-microscopic droplets become hail. Hail falls from the sky once it has grown so heavy that the updraft can no longer keep it suspended. The stronger a storm's updrafts, the longer the hail can remain suspended, tumbling and growing larger. Weaker updrafts can support hail, but the stones are much smaller.

Extreme Hailstorms

Large hail streak: On June 26, 1968, an Illinois hailstorm pelted a region 19 miles wide and 51 miles long with an estimated 82 million cubic feet of ice.

The costliest hailstorm in U.S. history: On May 5, 1995, a hailstorm pummeled Fort Worth and Tarrant Counties in Texas, near Dallas. Damage from hail and severe flash flooding caused by drifted hail totaled over $1.2 billion. The world's costliest hailstorm hit Munich Germany in 1984, costing $1 billion American ($1.7 billion in 2002 dollars).

The second-costliest hailstorm in U.S. history: On July 11, 1990, a severe hailstorm pounded the Front Range of Colorado. Property damages in Denver soared over $625 million ($860 million in 2002 dollars).

The Butte-Meade storm: On July 5, 1996, a catastrophic storm pushed a 70-mile-long, 8-mile-wide hail swath through northwestern South Dakota in Butte and Meade Counties. The storm's 1- to 3-inch-diameter hail and 100-plus-mile-per-hour winds caused a scar 100 miles long and 4 to 5 miles wide. The scar was so large that is was visible from satellites for a month. It killed scores of livestock, including fifty-six horses at a single ranch, and destroyed 200,000 acres of private rangeland while damaging another 300,000 acres.

The Rocky Mountain Front and Great Plains are prone to hail, also called "ice fruit" or "lumpy rain," because dry continental air masses clash with humid subtropical air masses in this region and produce thunderstorms. In addition, the thin atmosphere over the relatively higher elevations of the Great Plains warms quickly under intense summer sunshine. This enhances thermal instability, which causes rapid updrafts and thunderheads. At these higher elevations hail doesn't have to fall as far before reaching the ground either. The short trip reduces the chance the hail will melt. Central Florida—at or near sea level—receives more than one hundred thunderstorms each year, but it has a much lower incidence of hail. Hail melts to large raindrops before reaching the surface.

Hail often falls in swaths or streaks beneath its parent thunderstorm. Since the mature stage of a hail-producing thunderstorm lasts only thirty minutes or so, the typical hail streak is 5 miles long and ½ mile wide. However, slow-moving lines of severe thunderstorms can produce much longer and wider hail streaks, because the thunderstorms grow in size and power with time. The largest hail streaks stretch over 70 miles long and 20 miles wide. A slow-moving hailstorm can quickly turn a green, summery countryside into a white, wintry-looking landscape. Hailstorms typically last for one to six minutes, accumulating very little. Hail that continues for more than fifteen minutes is unusual, thirty minutes is rare, and at that point it has time to collect to appreciable depths. Although infrequent, it is not out of the ordinary for snowplows to be called out to clear hail from roads and highways in Hail Alley.

Hailstones measuring 3 inches or greater in diameter are rare. Most hailstones range in size from that of a pea to a golf ball. Scientists have conducted few studies regarding North American hailstones, but they have found that 95 percent of North American hailstones measure ½ inch or less in diameter. Hail ¾ inch or greater in diameter can damage structures. The National

Hard Hat Area

A pea-sized hailstone can reach 18 miles per hour before hitting the ground. A golf ball–sized stone can fall at 56 miles per hour, while a grapefruit-sized (4 inches in diameter) hailstone can reach nearly 150 miles per hour with enough energy to punch holes through the roofs of homes and cars.

A hailstone that landed on Potter, Nebraska, on July 6, 1928, was the largest hailstone recorded in the United States for a while. It was 5.5 inches in diameter, had a circumference of 17 inches, and weighed 1.5 pounds. On September 3, 1970, however, a severe thunderstorm dropped a 1.67-pound hailstone measuring 17.5 inches in diameter on Coffeyville, Kansas, making it the nation's largest official hailstone to date. Then in 2003 a cantaloupe-sized hailstone landed in Aurora, Nebraska, on June 22. It measured 7 inches in diameter, had a circumference of 18.75 inches, and weighed just under 1 pound—the largest hailstone ever recorded in the United States. The National Weather Service reported that the Aurora hailstone didn't break the record for the heaviest hailstone. The NWS could not get an accurate weight because the hailstone hit a rain gutter on its descent, losing about 40 percent of its mass. (The measurements were of the remaining, broken hailstone.) A severe thunderstorm produced this hailstone along with other hail big enough to punch through the roofs of homes!

This hailstone, the heaviest hailstone documented in the United States, weighed 1.67 pounds and fell in Coffeyville, Kansas, in 1970. Similar to tree rings, this hailstone's rings show how it grew. —Photo courtesy of the University Corporation for Atmospheric Research (UCAR)

Weather Service calls thunderstorms that spawn hail this large *severe thunderstorms*.

While any size hailstone can devastate crops, larger stones can harm livestock and people. Two known hail-related deaths have occurred in the United States. On July 30, 1979, a 3-inch-diameter hailstone struck and killed an infant in Fort Collins, Colorado. In May 1930, a farmer caught in an open field near Lubbock, Texas, died as a result of a violent hailstorm. While humans can usually escape, hailstorms kill a great number of livestock each year as the animals graze in open fields.

Depending on wind speeds, duration of the hailstorm, and hailstone sizes, the destruction within hail swaths can range from minor to catastrophic. In the most catastrophic cases, the continual barrage of wind-driven hail kills rangeland grasses, completely defoliates and debarks trees, snaps tree branches, exposes tree roots, strips the paint off buildings, or pounds the

siding off altogether. In some cases hailstones can even become lodged in siding. Insects and birds die, leaving a swath eerily silent after a storm. A Great Plains hailstorm can reap damage as severe as a weak to moderate tornado.

Hail season in Hail Alley runs from March to October, but the majority of hail-producing storms occur between May and July, when thunderstorms occur most frequently. Hail, primarily an afternoon and evening phenomenon, occurs after the sun has heated the earth's surface for a while; nearly 90 percent of all severe hailstorms occur between 1:00 and 9:00 PM.

Throughout the centuries and around the world, people have tried to prevent hail. In the fourteenth century, Europeans attempted to ward off hail by ringing church bells and firing "hail cannons" into thunderstorms. Not only did they produce an appalling sound, but also they sent a smoke ring ascending into the clouds at almost 100 feet per second. People believed that the shock

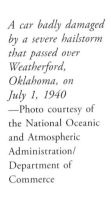

A car badly damaged by a severe hailstorm that passed over Weatherford, Oklahoma, on July 1, 1940
—Photo courtesy of the National Oceanic and Atmospheric Administration/ Department of Commerce

waves from the hail cannon would break up hail particles, creating sleet or rain, and the smoke would provide nuclei for condensation to collect on, allowing rain to form before a storm grew big enough to produce hail. Especially famous in the wine-producing regions of Europe during the nineteenth century, these cannons and modern versions of them are still used in parts of Italy.

One of the larger hail-suppression efforts, which involved cloud seeding, took place in North Dakota in the 1960s. Farmers looking to decrease their losses from hail paid for this project. Hail usually forms in high-level clouds (above 30,000 feet), but people generally seed clouds at 20,000 feet. Scientists speculated that by seeding storms before they grew large, they could force storms to rain themselves out before they could produce hail. Once a storm cloud grows too tall, seeding is impossible and even dangerous because of the intense turbulence caused by a thunderstorm's updraft and downdraft. Although some studies indicate that seeding does lead to less hail and more precipitation, other studies remain inconclusive. This lack of conclusion has limited funding for additional hail-suppression research. No reliable defense against hail exists. Purchasing insurance is about the only thing farmers can do to protect themselves from the cost of hail damage.

Blizzards

The rolling prairies that make up the Dakotas offer little resistance to high winds, drifting snow, and frigid temperatures. Even the Black Hills region, which presents some resistance to wind, is not immune to blizzard conditions. Blizzards in the Black Hills, often less severe than elsewhere in the Dakotas, can still produce heavy drifting snow. South Dakota has the dubious distinction of being called the Blizzard State, since it has the highest frequency of blizzards of any state in the lower forty-eight. Blizzards can occur as far south as Texas, which is also known as the Blizzard State, and as far east as Maine.

In order for a storm to become a blizzard, according to the U.S. Weather Bureau's (now called the National Weather Service) official definition of 1958, winds must exceed 35 miles per hour with sufficient snow in the air to reduce visibility to less than ¼ mile for a period of three hours or more. An official blizzard does not require falling snow, just enough existing snow to blow into a whiteout. Earlier definitions included temperatures in the 10 to 20°F range for a severe blizzard, but meteorologists no longer use this criterion. The word *blizzard* originated in the United States, but is also used in other countries. Coined on March 14, 1870, the term described a storm that produced heavy snow and high winds in Minnesota and Iowa.

On average, the Dakotas experience two to three blizzards each year; as a whole, the United States averages eleven blizzards annually. During the winter of 1996–97, though, the Dakotas experienced nine blizzards. Blizzards in the Great Plains generally result from strong low-pressure systems moving eastward or northeastward on the lee side of the Rocky Mountains. Dakota blizzards often form as the result of two different, yet common, low-pressure systems: Alberta lows, which form in southern Alberta, and Colorado lows, which form over southern Colorado (see *chapter 11*).

About thirty-nine Colorado lows move across the plains during fall and spring, and in an average year one or two develop into blizzards in the Dakotas. About forty-two Alberta lows develop during winter, but usually only one of them intensifies enough to cause blizzards in North Dakota. Alberta lows tend to move fast since they ride the polar jet stream, and their blizzard conditions generally don't last as long as those caused by Colorado lows. However, Alberta blizzards pose more of a threat to life because the blizzard conditions often develop rapidly along with subzero temperatures. Since Colorado lows form and move into the Dakotas from the south,

A ground blizzard partially obscures a roadway in northeastern Colorado.

Ground Blizzards

Like a white veil, a ground blizzard can conceal the surface with a thin sheet of blowing snow on an otherwise sunny day. The Great Plains from northeastern Colorado to North Dakota commonly gets hit by frustrating and dangerous ground blizzards. Unlike blowing snow, which reduces visibility 6 feet or more above the surface, ground blizzards—referred to as *drifting snow* by weather forecasters—hug the ground and don't blow more than a foot above it. Ground blizzards are completely opaque, covering the surface in an undulating layer of snow. The most annoying ground blizzards strike on clear days, when motorists expect decent driving conditions. Motorists that attempt to drive through a ground blizzard often find themselves stranded off the road with no idea where the road went.

Both blizzards and ground blizzards make driving on the Great Plains risky. Within three months of its opening in October 1970, the 77-mile stretch of Interstate 80 between Laramie and Walcott Junction, Wyoming, closed for ten days because winds up to 100 miles per hour created deep snowdrifts and near-zero visibility. Sixteen-foot-tall snowdrifts buried the road in twenty-seven places. Frustrated by massive drifts, the Wyoming Department of Transportation installed snow fences to protect motorists from future storms and drifting. Research found plowing snow one hundred times more expensive than intercepting it. So today, drivers can see miles and miles of snow fences lining freeways and roads throughout the Great Plains.

A snow fence reduces wind speed and increases turbulence in an air mass, causing snow to fall to the ground rather than continuing to blow horizontally. Snow collects on the lee side of the snow fence. During ground blizzards snow fences often provide a narrow clearing for motorists to see the road. If not for these fences, Interstate 80 would close more often during winter.

Forty-Foot Snowdrifts

The worst blizzard on record, in terms of the amount of time it lasted, blasted the northern Great Plains between March 2 and 4, 1966. Broken Bow, Nebraska, recorded wind gusts in excess of 100 miles per hour and snow drifts 40 feet tall. Bismarck, North Dakota, had near-zero visibility for forty-two consecutive hours and received 22.4 inches of snow, setting a new single-storm snowfall record for North Dakota. Mobridge, South Dakota, recorded 35 inches. Thirteen people died as a result of the blizzard, and livestock losses were heavy.

they tend to be warmer and wetter since they've collected Gulf of Mexico moisture along their journey. They tend to produce more snow than their drier, cooler Alberta counterparts. About one-eighth of the blizzards in the Dakotas are created by storm systems other than Alberta or Colorado lows, such as rapidly moving cold fronts or different low-pressure systems.

Blizzards caused by Colorado lows most commonly strike in November and March, while Alberta lows bring blizzards between December and February. From December to February, the Canadian storm track that directs Alberta lows sits at its farthest southern position, affecting the plains along the Canadian border more frequently than in other months. In November and March, the polar jet stream, which spawns

*Utility poles on March 9, 1966, near Jamestown, North Dakota, after the famous March 2–4, 1966, blizzard that buried the northern Great Plains —*Bill Koch photo, courtesy of the Collection of Dr. Herbert Kroehl, NGDC

Colorado lows, shifts northward and therefore affects the Dakotas more frequently than in the other months.

The strong winds associated with Dakota blizzards result from the strong pressure gradient that develops between high pressure in Canada and low pressure associated with either a Colorado low or an Alberta low. The counterclockwise flow around these lows causes easterly—shifting to northerly—winds across the Dakotas during blizzards.

Blizzard conditions in the Dakotas seldom last more than a day or two and frequently only a few hours. Most Dakotans respect blizzards and think of them as minor and temporary inconveniences to their normal lifestyle.

Snow Rollers

Though rare, snow rollers occur when strong winds couple with just the right snow conditions. For them to occur the temperature must hover just above freezing, and a fresh layer of snow must cover an old, harder layer of snow. Gusts of wind can grab the fresh snow and roll it across a field. With enough momentum, these rollers can reach diameters of over 2 feet.

A snow roller —Photo courtesy of the University Corporation for Atmospheric Research (UCAR)

References

Books and Journal Articles

Ahrens, C. D. 1994. *Meteorology Today: An Introduction to Weather, Climate, and the Environment.* St. Paul, Minn.: West Publishing Company.

Ashcroft, G. L., D. T. Jensen, and J. L. Brown. 1992. *Utah Climate.* Logan, Utah: Utah Climate Center.

Avery, C. C., L. R. Dexter, R. R. Wier, W. G. Delinger, A. Tecle, and R. J. Becker. 1992. Where has all the snow gone? Snowpack sublimation in northern Arizona. *Proceedings of the 60th Western Snow Conference,* Jackson Hole, Wyo., 84–92.

Bates, G. T. 1990. A case study of the effects of topography on cyclone development in the western United States. *Monthly Weather Review* 118:1808–25.

Bentley, W. A., and W. J. Humphreys. 1962. *Snow Crystals.* New York: Dover Publications.

Bettge, T. W. 1985. The case of the Bennett, Colorado, maximum temperature. *Weatherwise* 38:95–97.

Bluestein, H. B. 1992. *Synoptic-Dynamic Meteorology in Midlatitudes.* Vol. 1, *Principles of Kinematics and Dynamics.* Vol. 2, *Observations and Theory of Weather Systems.* New York: Oxford University Press.

Buehner, T. 1998. 1997: A record year for tornadoes in Washington. Presented at the Pacific Northwest Weather Workshop, Seattle, Wash., February 20–21, 1998.

Burroughs, W. J., B. Crowder, T. Robertson, E. Vallier-Talbot, and R. Whitaker. 1996. *Weather.* Singapore: Kyodo Printing Company.

Carr, M. 2000. Digging out…or digging in? *Northwest Science and Technology* Winter: 18–23.

Changnon, S. A. 2001. *Thunderstorms Across the Nation: An Atlas of Storms, Hail, and Their Damages in the 20th Century.* Champaign, Ill.: Printec Press.

Chaston, P. R. 1996. *Hurricanes!* Kearney, Mo.: Chaston Scientific.

Colle, B. A., and C. F. Mass. 1998. Windstorms along the western side of the Washington Cascade Mountains. Pt. 2, Characteristics of past events and three-dimensional idealized simulations. *Monthly Weather Review* 126:53–71.

Cotton, W. R. 1990. *Storms.* Fort Collins, Colo.: *ASTeR Press.

Crook, N. A., T. L. Clark, and M. W. Moncrieff. 1990. The Denver cyclone. Pt. 1, Generation in low Froude number flow. *Journal of the Atmospheric Sciences* 47(23):2725–42.

Daly, C., G. Taylor, and W. Gibson. 1997. The PRISM approach to mapping precipitation and temperature. In *Proceedings of the 10th American Meteorological Society Conference on Applied Climatology,* Reno, Nev., October 20–23, 1997, 10–12.

Daseler, J. Hailstorms at Cheyenne, Wyoming, 1892–1996. NOAA Technical Memorandum CR-114. National Weather Service Forecast Office.

De Blij, H. J. 1994. *Nature on the Rampage.* Washington, D.C.: Smithsonian Books.

DeVoir, G. 1998. Arctic air in the northern inter-mountain region: A study of climatologically favorable upper air patterns from the Columbia Basin to the Snake River plain. Presented at the Pacific Northwest Weather Workshop, Seattle, Wash., February 21–21, 1998.

Doesken, N. J. 1994. Hail, hail, hail—the summertime hazard of eastern Colorado. *Colorado Climate* 17(7):1–10.

———. 2000. The winter cloud of the Rockies. *Colorado Climate* 2(1):1.

Doesken, N. J., and A. Judson. 1996. *The Snow Booklet: A Guide to the Science, Climatology, and Measurement of Snow in the United States.* Fort Collins, Colo.: Department of Atmospheric Science, Colorado State University.

Doesken, N. J., and T. B. McKee. 1997a. Colorado extreme storm precipitation data study. *Climatology Report,* no. 97-1. Department of Atmospheric Science, Colorado State University, Fort Collins, Colo.

———. 1997b. The Silver Lake, Colorado, snowstorm of April 14–15, 1921. In *Proceedings of the 10th Conference on Applied Climatology,* Reno, Nev., October 20–23, 1997, 153–56.

Doty, S. 2004. A true weather observing hero. *The Climate Station Chronicles* 5:1–3.

Eagleman, J. R. 1990. *Severe and Unusual Weather*. Lenexa, Kans.: Trimedia Publishing Company.

Espenshade, E. B., Jr., ed. 1990. *Goode's World Atlas*. Chicago, Ill.: Rand McNally.

Fujita, T. T. 1989. The Teton-Yellowstone Tornado of 21 July 1987. *Monthly Weather Review* 117:1913–40.

Gibilisco, S. 1984. *Violent Weather: Hurricanes, Tornadoes and Storms*. Blue Ridge Summit, Pa.: Tab Books.

Glickman, T. S., ed. 2000. *Glossary of Meteorology*. 2nd ed. Boston, Mass.: American Meteorological Society.

Goldstein, M. 1999. *The Complete Idiot's Guide to Weather*. New York: Alpha Books.

Haynes, A. 2001. Synoptic pattern typing for historical heavy precipitation events in southern California. Technical Attachment no. 1-15, National Oceanic and Atmospheric Administration.

Henson, R. 2002. *The Rough Guide to Weather*. London: Rough Guides Ltd.

Hickcox, D. H. 2001. Temperatures extremes. *Weatherwise* 54:50–56.

Higgins, S. O., and L. B. Lee, eds. 1996. *Ice Storm '96: Days of Darkness, Days of Cold*. Spokane, Wash.: Lawton Printing.

Horvitz, A. A national temperature record at Loma, Montana. Presented at the American Meteorological Society 12th Symposium on Meteorological Observations and Instrumentation, Silver Spring, Md. February 8–13, 2003.

Jennings, A. H. 1952. Maximum 24-hour precipitation in the United States. Technical Paper no. 16, U.S. Department of Commerce, Weather Bureau, Washington, D.C.

Keen, R. A. 1987. *Skywatch: The Western Weather Guide*. Golden, Colo.: Fulcrum.

Kimerling, A. J., and P. L. Jackson, eds. 1985. *Atlas of the Pacific Northwest*. Corvallis, Oreg.: Oregon State University Press.

Koss, W. J., J. R. Owenby, P. M. Steurer, and D. S. Ezell. 1988. Freeze/frost data. *Climatography of the U.S.* 20: S1. Asheville, N.C.: National Climatic Data Center.

Laskin, D. 1996. *Braving the Elements: The Stormy History of American Weather*. New York: Doubleday.

Leffler, R. J., A. Horvitz, R. Downs, M. Changery, K. T. Redmond, and G. Taylor. 2001. Evaluation of a national seasonal snowfall record at the Mount Baker, Washington, ski area. *National Weather Digest* 25(1,2):15–20.

Lindley, T. 1997. Effects of Texas panhandle topography on dryline movement. Technical Attachment no. SR/SSD 97-39, National Weather Service, Amarillo, Tex.

Lockhart, G., 1988. *The Weather Companion: An Album of Meteorological History, Science, and Folklore*. New York: Dodd, Mead & Company.

Ludlum, D. M. 1982. *The American Weather Book*. Boston, Mass.: Houghton Mifflin Company.

———. 1995. *National Audubon Society Field Guide to North American Weather*. New York: Alfred A. Knopf.

Lynch, D. K. 1980. *Atmospheric Phenomena*. San Francisco: W. H. Freeman & Company.

Lynott, R. E., and O. P. Cramer. 1966. Detailed analysis of the 1962 Columbus Day windstorm in Oregon and Washington. *Monthly Weather Review* 94:105–17.

Lyons, W. A. 1996. *The Handy Weather Answer Book*. Detroit, Mich.: Visible Ink Press.

McKee, T. B., and N. J. Doesken. 1998. An analysis of rainfall for the July 28, 1997, flood in Fort Collins, Colorado. *Climatology Report*, no. 98-1, Department of Atmospheric Science. Colorado State University, Fort Collins, Colo.

Miller, J. F., R. H. Frederick, and R. J. Tracey. 1973. Precipitation-frequency atlas of the western United States. *NOAA Atlas* 2, U.S. Department of Commerce, National Oceanic and Atmospheric Administration, National Weather Service, Silver Spring, Md.

Miller, R. J., and B. Bernstein. 1998. A climatology of freezing precipitation in the Columbia Basin. Presented at the Pacific Northwest Weather Workshop, February 20–21, 1998, Seattle, Wash.

Nair, U. S., M. R. Hjelmfelt, and R. A. Pielke, Sr. 1997. Numerical simulation of the June 9–10, 1972, Black Hills storm using CSU RAMS. *Monthly Weather Review* 125:1753–66.

National Oceanic and Atmospheric Administration. 1880–2004. *Climatological Data*. Since I referred to so many of these it is impractical to list them all. This government publication is released monthly. Provided you have the date and state of the particular event, you can easily reference this publication for detailed climate information. In earlier volumes, descriptive narratives of storm events were particularly helpful in describing notable events in this book.

———. 1959–2004. *Storm Data*. Since I referred to so many of these it is impractical to list them all. This government publication is released monthly. Each issue is broken down by state and day. Provided you have the date and state of the particular event, you can easily reference the appropriate storm data for information.

————. 1974. *Weekly Weather and Crop Bulletin* 61 (3–4).

————. 1982. Climate of Arizona. *Climatography of the United States,* no. 60. I used this same issue for climate information regarding California, Colorado, Idaho, Kansas, Montana, Nebraska, Nevada, New Mexico, North Dakota, Oklahoma, Oregon, South Dakota, Texas, Utah, Washington, and Wyoming.

Nelson, M. 1999. *The Colorado Weather Book.* Englewood, Colo.: Westcliffe Publishers.

Neumann, C. J., B. R. Jarvinen, C. J. McAdie, and J. D. Elms. 1993. *Tropical Cyclones of the North Atlantic Ocean, 1871–1992.* Historical Climatological Series 6–2. Asheville, N.C.: National Climatic Data Center.

Quayle, L. 1990. *Weather: Understanding the Forces of Nature.* New York: Crescent Books.

Ruffner, J. A., and F. E. Bair, eds. 1977. *The Weather Almanac.* New York: Avon Books.

Sanders, T. 1983. *The Weather Is Front Page News.* South Bend, Ind.: Icarus.

Schaefer, V. J., and J. A. Day. 1981. *A Field Guide to the Atmosphere.* Boston, Mass.: Houghton Mifflin Company.

Schroeder, M. J., and C. C. Buck. 1970. Fire weather: A guide for application of meteorological information to forest fire control operations. Agricultural Handbook 360, U.S. Department of Agriculture, Forest Service.

Steenburgh, W. J., S. F. Halvorson, and D. J. Onton. 2000. Climatology of lake-effect snowstorms of the Great Salt Lake. *Monthly Weather Review* 128:709–27.

Szoke, E. J., and B. L. Shaw. 2001. An examination of the operational predictability of mesoscale terrain-induced features in eastern Colorado from several models. In *Proceedings of the Ninth Conference on Mesoscale Processes,* Fort Lauderdale, Fla., July 30–August 2, 2001.

Taylor, G. H., and R. R. Hatton. 1999. *The Oregon Weather Book.* Corvallis, Oreg.: Oregon State University Press.

Thompson, W. T., S. D. Burk, and J. Rosenthal. 1997. An investigation of the Catalina Eddy. *Monthly Weather Review* 125:1135–46.

Tufty, B. 1987. *1001 Questions Answered about Hurricanes, Tornadoes and Other Natural Air Disasters.* New York: Dover Publications.

Uman, M. A. 1986. *All About Lightning.* New York: Dover Publications.

Watson, B. A. 1993. *The Old Farmer's Almanac Book of Weather and Natural Disasters.* New York: Random House.

Williams, J. 1992. *The Weather Book: An Easy-to-Understand Guide to the USA's Weather.* New York: Vintage Books.

————. 1994. *The USA Today Weather Almanac 1995.* New York: Vintage Books.

Woolley, R. R. 1946. Cloudburst floods in Utah, 1850–1938. Water-Supply Paper no. 994, U.S. Geological Survey. Washington, D.C.: U.S. Government Printing Office.

Web Sites

Climate Source, Inc. Latest PRISM data. http://www.climatesource.com/ (accessed April 2004).

Curtis, Jan, and Kate Grimes. Wyoming climate atlas. Water Resources Data System, Wyoming State Climate Office. http://www.wrds.uwyo.edu/wrds/wsc/ climateatlas/toc.html (accessed April 2004).

High Plains Regional Climate Center. University of Nebraska, Lincoln. http://www.hprcc.unl.edu/ (accessed April 2004).

Hydrometeorological Design Studies Center. Record rainfalls. NOAA National Weather Service. http://www.nws.noaa.gov/oh/hdsc/max_precip/ max_precip.htm (accessed April 2004).

Miller, Miguel. The weather guide: A weather information companion for the forecast area of the National Weather Service in San Diego. January 2004. NOAA National Weather Service. http://www.wrh.noaa.gov/sandiego/research/ Guide /The_Weather_Guide.pdf (accessed April 2004).

National Environmental Satellite, Data, and Information Service. NOAA satellites and information. NOAA National Climatic Data Center. http://www.ncdc.noaa.gov/oa/ncdc.html (accessed April 2004).

National Environmental Satellite, Data, and Information Service. State climate offices. National Climatic Data Center. http://www.ncdc.noaa.gov/oa/climate/ stateclimatologists.html (accessed April 2004).

National Interagency Fire Center. http://www.nifc.gov/ (accessed April 2004).

National Oceanic and Atmospheric Administration. Climate atlas of the United States. NOAA National Climatic Data Center. http://lwf.ncdc.noaa.gov/oa/about/cdrom/climatls2/atlashelp.html (accessed April 2004).

National Oceanic and Atmospheric Administration. Extreme weather and climate events. NOAA National Climatic Data Center. http://www.ncdc.noaa.gov/oa/climate/severeweather/extremes.html (accessed April 2004).

National Oceanic and Atmospheric Administration. U.S. daily weather maps project. NOAA Central Library. http://docs.lib.noaa.gov/rescue/dwm/data_rescue_daily_weather_maps.html (accessed April 2004).

National Oceanic and Atmospheric Administration and the Cooperative Institute for Research in Environmental Sciences. Climate diagnostics center. http://www.cdc.noaa.gov/ (accessed April 2004).

National Weather Service. National Oceanic and Atmospheric Administration. http://www.nws.noaa.gov/ (accessed April 2004).

National Weather Service. Storm prediction center. National Oceanic and Atmospheric Administration. http://www.spc.noaa.gov/ (accessed April 2004).

National Weather Service. Weather events of the century by state. National Oceanic and Atmospheric Administration. http://www.wrh.noaa.gov/topweather/ (accessed April 2004).

Natural Resources Conservation Service. National water and climate center. U.S. Department of Agriculture. http://www.wcc.nrcs.usda.gov/ (accessed April 2004).

Plymouth State Weather Center. Make your own product generator for archived data. Plymouth State University. http://vortex.plymouth.edu/u-make.html (accessed April 2004).

Read, Wolf. The storm king: Some historical weather events in the Pacific Northwest. Oregon Climate Service, Oregon State University. http://oregonstate.edu/~readw/index.html (accessed April 2004).

Spatial Climate Analysis Service. Latest PRISM data. Oregon Climate Service, Oregon State University. http://www.ocs.orst.edu/prism/ (accessed April 2004).

U.S. Geological Survey. U.S. Department of the Interior. http://www.usgs.gov/ (accessed April 2004).

Utah Center for Climate and Weather. Major Utah weather events, 1847–1995. http://www.utahweather.org/UWC/utahs_daily_wx_almanac/1847_1995.html (accessed April 2004).

The Weather Channel. http://www.weather.com/ (accessed April 2004).

Western Regional Climate Center. Historical climate information. http://www.wrcc.dri.edu CLIMATEDATA.html (accessed April 2004).

Western Regional Climate Center. Snotel graph. Desert Research Institute. http://www.wrcc.dri.edu/cgi-bin/snoX_graphX.pl (accessed April 2004).

Williams, Jack. Ask Jack questions and answers archives. USA Today Weather. http://www.usatoday.com/weather/resources/askjack/wjack5.htm (accessed April 2004).

Correspondence

In researching this book, my correspondence with a number of professional acquaintances influenced what I wrote. These include:

Doesken, Nolan. Assistant State Climatologist, Colorado. (Hail Alley; Champagne Powder and World Record Snow Intensity: Colorado.)

Kelly, Robert. Seattle, Wash. (The Wellington Avalanche: Stevens Pass, Washington.)

Potter, Sean. National Weather Service. (Daily Temperature Extremes: Arizona and California.)

Redmond, Kelly. Western Regional Climatologist. (Dangerous Snowstorms.)

Riley, David. National Weather Service. (The Marfa Dryline: Western Texas; Tornado Alley.)

Taylor, George. Oregon State Climatologist. (The Pineapple Express and the Flood of '96, Thermal Low: Willamette Valley; The Columbus Day Storm.)

Wing, Warren. (The Wellington Avalanche: Stevens Pass, Washington.)

Glossary

These definitions have been modified from the Weather Channel's Web site, weather.com, and from the American Meteorological Society's *Glossary of Meteorology*.

adiabatic process. A process of heat transfer within a system (such as an air parcel). In the adiabatic process, heat is not transferred to anything that surrounds the system, and compression of molecules always results in warming, while expansion results in cooling; for example, the temperature of descending air, such as air flowing down the slopes of mountains, can increase 5.5°F for every 1,000 feet of descent.

advection. The horizontal transfer of an air mass by the wind. For example, warm air advection occurs when winds move warm air into a region, in contrast to air within a region that warms on its own.

advection fog. Occurs when warm, moist air moves over a cold surface and cools below its dew point.

air density. Mass per unit volume of air; for example, 1.275 kg per cubic meter at 0°C and 1,000 millibars of pressure. The density of air fluctuates with temperature, elevation, and moisture content.

air mass. A large parcel of air that has similar temperature and moisture content throughout.

air pressure. The cumulative downward force, expressed as the weight of a column of air per unit of surface area, exerted on any surface by air molecules. Atmospheric pressure is the force exerted against the earth's surface by the weight of the air above that surface.

aloft. In the upper atmosphere.

altocumulus. A mid-atmosphere cloud, typically between 6,500 and 20,000 feet, that is usually white or gray. Often occurs in layers or patches with wavy, rounded masses or rolls.

ambient air. Air surrounding a cloud or rising or sinking air parcel that doesn't have the same atmospheric characteristics as the air parcel it surrounds.

anemometer. An instrument used to measure wind speed.

anomalies. In terms of climate, departures in temperature, precipitation, or other weather elements from long-term (thirty-year) averages.

anticyclone. A dome of air with relatively higher atmospheric pressure than the surrounding atmosphere. In the northern hemisphere, winds in anticyclones blow clockwise; counterclockwise in the southern hemisphere. Also known as a *high* or *ridge*.

arctic air. A very cold and dry air mass that primarily forms in winter and in the northern interior of North America. It often moves into, and affects the weather of, the United States.

atmospheric pressure. See **air pressure.**

barometer. An instrument that measures atmospheric pressure. The two most common barometers are the aneroid barometer and the mercurial barometer. An aneroid barometer contains no liquid and measures pressure in millibars. A mercurial barometer contains mercury and measures pressure in inches as mercury rises and drops in a glass tube.

blizzard. A severe weather condition, characterized by low temperatures and strong winds (greater than 32 miles per hour), that causes a great deal of blowing snow and frequently reduces the visibility to ¼ mile or less for at least three hours. When these conditions persist after snow has ceased falling, meteorologists term the event a *ground blizzard.*

bora. A cold katabatic wind that originates in Yugoslavia and flows onto the coastal plain of the Adriatic Sea. The term is also used to describe any cold wind flowing from higher to lower terrain, such as from the Rocky Mountains to the Great Plains.

Chinook. A warm, dry foehn that flows east off the Rocky Mountains. These winds warm through the adiabatic process, and they can drastically raise temperatures in areas on the lee side of the Rockies.

cirrus. A high cloud composed of ice crystals. They appear as thin, white, featherlike clouds in patches, filaments, or narrow bands.

cloud base. The lowest portion of a cloud.

cloudburst. A sudden and heavy rain shower.

cloud seeding. The introduction of artificial substances (usually silver iodide or dry ice) into a cloud for the purpose of either modifying its development or increasing its precipitation.

cold air funnels. Also known as *cold core funnels,* these short-lived vortexes develop from relatively small showers or nonsupercell thunderstorms when the air aloft is very cold. They usually stay aloft, but sometimes touch down briefly.

cold front. The leading edge of a cold air mass, where a lot of stormy weather occurs.

compressional heating. Heating often associated with adiabatic compression. See also **adiabatic process**.

condensation. Process by which water changes phase from a vapor to a liquid.

condensation nuclei. Small particles in the atmosphere on which water vapor condenses to become rain, hail, or snow. The nuclei may be dust, salt, or other materials.

convection. The motions in a fluid, like air, that result in the transport and mixing of the fluid's properties, such as temperature. Convection predominantly drives vertical atmospheric motions, such as rising air currents caused by surface heating. The rising of heated surface air and the sinking of cooler air aloft is often called *free convection,* as opposed to *forced convection,* which is air forced upward by topography.

convective storm. A thunderstorm caused by localized convection.

convergence. An atmospheric condition that exists when horizontal winds collide in a particular region. Contrast **divergence**.

Coriolis effect. A deflective force resulting from the rotation of the earth on its axis; it principally affects large-scale and global-scale winds. Winds are deflected to the right of their initial direction in the northern hemisphere, and to the left in the southern hemisphere.

cumulonimbus. An extremely dense and vertically developed cloud, often with a top in the shape of an anvil. These clouds are accompanied by heavy showers, lightning, thunder, and sometimes hail. They are also known as a *thunderstorm clouds.*

cumulus. A cloud in the form of individual, detached domes or towers that are usually dense and well defined. It has a flat base with a bulging upper part that often resembles cauliflower.

A cumulus cloud of fair weather is called *cumulus humilis;* one that exhibits a great deal of vertical growth is called *towering cumulus.*

cutoff high. A high-pressure system that separates from the prevailing westerly airflow and remains stationary, or cut off from the main flow.

cutoff low. A low-pressure system that separates from the prevailing westerly airflow and remains stationary, or cut off from the main flow.

cyclone. A surface low-pressure system with counterclockwise wind circulation in the northern hemisphere. Stormy weather is characteristic of cyclones.

density. The ratio of the mass of a substance to the volume it occupies.

deposition. The process by which water vapor in subfreezing air changes directly to ice without first becoming a liquid. See also **sublimation**.

dew. Water vapor that has condensed on objects near the ground. When the temperatures of objects near the ground are lower than the dew point of an air mass, condensation occurs.

dew point (or dew-point temperature). The temperature to which air must be cooled (at a constant pressure and with a constant water vapor content) for it to be saturated with moisture. When the dew point falls below freezing it is called the *frost point.*

divergence. An atmospheric condition that exists when horizontal winds expand or spread out from a region. Contrast **convergence**.

downburst. A severe localized downdraft greater than 2½ miles across that occurs beneath a severe thunderstorm. See also **microburst**.

downdraft. Downward moving air, usually within a thunderstorm cell.

drizzle. Small drops of moisture between 0.2 and 0.5 millimeters in diameter that fall slowly and re-

duce visibility more than light rain. Sometimes called *mist.*

dry adiabatic rate. The rate of change of temperature in a rising or descending unsaturated (dry) air parcel. The rate of adiabatic cooling or warming is 5.5°F per 1,000 feet.

dryline. A boundary between warm, dry desert air from the southwestern United States and warm, moist air from the Gulf of Mexico. It usually spans north-south across the central and southern Great Plains during the spring and summer and is an area of thunderstorm activity.

dust devil. A small but rapidly rotating wind (up to 60 miles per hour) made visible by the dust, sand, and debris it picks up from the surface. Also called *whirlwinds,* they develop on clear, dry, hot afternoons.

eddy. A relatively small volume of air that behaves differently from the larger flow in which it exists.

evaporation. The process by which a liquid changes into a gas.

evaporative cooling. The cooling of an air mass that occurs when water changes to its gaseous state (water vapor).

evapotranspiration. The total amount of water that is transferred from Earth's surface to the atmosphere by the evaporation of liquids and solids and transpiration by plants.

fall freeze date. The first date in fall when the temperature drops below freezing.

fetch. The surface area of water over which the wind travels at a constant speed and direction. In terms of lake effect snow, it is the surface area of a lake a storm gathers moisture from as the storm travels over it.

foehn. A warm, dry wind that descends the lee side of a mountain range; for example, a Chinook. Its temperature warms through compressional heating.

fog. A cloud base that occurs at the earth's surface and reduces visibility to less than ⅝ mile. Fog occurs when the air temperature and dew point are the same (or nearly so), and water droplets, which have attached to condensation nuclei, become suspended in the atmosphere.

freeze. Occurs when the surface air temperature remains below freezing over a large area for a sufficient enough time to damage certain agricultural crops. Freezes occur most often as cold air is advected into a region.

freezing rain, freezing drizzle. Rain or drizzle that falls in liquid form and then freezes upon striking cold objects or the ground. Both can produce a coating of ice on objects, which is called *glaze*.

front. The transition zone between two distinct air masses.

frost. A thin, fuzzy layer of ice crystals deposited on an object, such as a window, bridge, or blade of grass. Frost occurs when the air temperature falls below the frost point.

frost point. See **dew point**.

funnel cloud. A rotating conelike cloud that extends down from the base of a thunderstorm. When it reaches the surface it is called a *tornado*.

glaze. A coating of transparent ice that forms when supercooled rain comes into contact with objects and freezes. A storm that produces glaze is called an *ice storm*.

ground fog. See **radiation fog**.

gust front. The leading edge of a cool outflow of air in a convective storm (often a supercell thunderstorm). Also called a *mesoscale cold front*.

haboob. A turbulent dust storm or sandstorm that forms as cold downdrafts from a thunderstorm lift dust and sand into the air.

hail. Solid precipitation; chunks or balls of ice with diameters greater than 5 millimeters.

hailstones. Transparent or partially opaque, pea-sized to golf ball–sized (or even larger) particles of ice that fall from cumulonimbus clouds.

haze. Fine, dry or wet dust or salt particles dispersed through a portion of the atmosphere. Individually they are not visible, but cumulatively they diminish visibility.

heat lightning. Distant lightning that illuminates the sky but is too far away for its thunder to be heard.

high-pressure system. An area of maximum atmospheric pressure, often depicted as an "H" on weather maps.

hoarfrost. Fernlike crystals of ice that form by deposition of water vapor on twigs, tree branches, and other objects.

hurricane. A severe tropical cyclone that has winds in excess of 73 miles per hour.

ice pellets. See **sleet**.

Indian summer. An unseasonably warm spell with clear skies that occurs after a period of significantly cold weather and often after the first freeze. It usually occurs near the middle of autumn.

insolation. The incoming solar radiation that reaches the atmosphere and the earth.

Intertropical Convergence Zone (ITCZ). The boundary separating the northeasterly trade winds of the northern hemisphere from the southeasterly trade winds of the southern hemisphere. It is characterized by a band of thunderstorms around the equator.

inversion. An increase in air temperature with altitude, which is a departure from the normal decrease in temperature with height.

isobar. A line connecting points of equal pressure on a weather map.

jet stream. A strong band of concentrated winds in the upper atmosphere. There is a jet stream in both the northern and southern hemisphere, and these currents steer storms. The paths jet streams follow are often called *storm tracks*.

katabatic wind. A wind that flows from high elevations down the slopes of mountains, plateaus, and hills to the valleys or plains below. In part, they are driven by gravity. Katabatic winds have many local names that relate to their relative temperature and location. For instance, the warm Chinook along the eastern slopes of the Rocky Mountains is a type of katabatic wind. See also **Chinook**.

lake breeze. A wind blowing onshore from the surface of a lake.

lake effect snow. Localized snowstorms that form downwind of lakes. Such storms are common in late fall and early winter near the Great Lakes and Great Salt Lake as cold, dry air picks up moisture and warmth from the unfrozen lakes.

land breeze. A coastal breeze that blows from land to sea usually at night, often resulting in warm land temperatures. Also known as an *offshore breeze*. Contrast **sea breeze**.

landspout. A small, weak tornado-like vortex that reaches the ground. Sometimes referred to as *nonsupercell tornadoes*, they are spawned from nonrotating convective storms.

lapse rate. The rate at which an atmospheric variable (usually temperature) decreases with altitude.

latent heat. The heat that is released or absorbed by a substance when it undergoes a change of state such as evaporation, condensation, or sublimation.

lenticular cloud. A lens-shaped cloud that forms on the lee side of mountains due to the wave wind pattern created by the mountains.

lightning. A visible electrical discharge produced by thunderstorms.

longwave radiation. A term most often used to describe the infrared energy emitted by the earth and the atmosphere. This type of radiation heats the air.

low-pressure system. An area of minimum atmospheric pressure, often depicted as an "L" on surface weather maps, and often indicating areas of stormy weather.

mammatus clouds. Clouds that look like pouches hanging from the underside of a cloud; they are usually associated with a strong thunderstorm.

mean annual temperature. The average temperature at any given location for the entire year. It is calculated by averaging the daily highs and lows.

mesoscale. The scale of meteorological phenomena that ranges in size from a few miles to about 62 miles. It includes local winds, thunderstorms, and tornadoes. Contrast **microscale**.

mesosphere. A thermal subdivision of the atmosphere where the temperature decreases with increasing altitude. The mesosphere, between 30 and 50 miles above the earth's surface, is situated above the stratosphere.

meteorology. The study of the atmosphere and atmospheric phenomena as well as the atmosphere's interaction with the earth's surface, oceans, and life in general.

microburst. A strong localized downdraft less than $2\frac{1}{2}$ miles wide that occurs beneath strong or severe thunderstorms. Contrast **downburst**.

microscale. The smallest scale of atmospheric motions, usually a mile or less. Contrast **mesoscale**.

millibar (mb). A unit for expressing atmospheric pressure. Sea level pressure is normally close to 1,013 mb.

mirage. A refraction phenomenon that makes an object appear to be displaced from its true position. When an object appears higher than it actually is, it is called a *superior mirage*. When an object appears lower than it actually is, it is an *inferior mirage*.

mist. Very light fog in which visibility is greater than 0.62 miles, but less than 5 miles.

moist adiabatic rate. The rate of change of temperature in a rising or descending saturated air parcel. The rate of cooling or warming varies, but a common value of 3.3°F per 1,000 feet is used.

monsoon. A seasonal wind pattern that usually brings a lot of moisture and rain to large areas.

mountain breeze, valley breeze. A local wind system of a mountain valley that blows downhill (mountain breeze) at night and uphill (valley breeze) during the day.

nocturnal inversion. An increase in temperature with altitude due to radiational cooling of the earth's surface. Also called *radiation inversion*.

offshore breeze. See **land breeze**.

onshore breeze. See **sea breeze**.

orographic lifting. The lifting of air over a topographic barrier. Clouds that form in this lifting process are called *orographic clouds*.

orographic precipitation. Rainfall or snowfall from clouds that formed as a result of orographic lifting.

polar air mass. A cold, dry air mass that forms in a high-latitude region.

potential evapotranspiration. The amount of moisture that, if it were available, would be removed from a given land area by evaporation and transpiration.

precipitation. Water particles in any form—liquid or solid—that fall from the atmosphere and reach the ground.

prevailing wind. The wind direction most frequently observed during a given period.

psychrometer. An instrument used to measure the water vapor content of the air.

radar. An instrument useful for remote sensing of meteorological phenomena. It operates by sending out radio waves and then monitoring those that are reflected back by objects such as raindrops within clouds.

radiational cooling. Cooling of the earth's surface and adjacent air, typically occurring on calm, clear nights when the longwave radiation emission from the surface exceeds shortwave or longwave radiation coming down from the atmosphere.

radiation fog. Fog produced over land when radiational cooling reduces the air temperature to or below its dew point. It is also known as *ground fog* and *valley fog*.

radiation inversion. See **nocturnal inversion**.

rain. Precipitation in the form of liquid water drops that have diameters greater than 0.5 millimeters.

rain shadow. The region on the lee side of a mountain where the precipitation is noticeably less than on the windward side.

relative humidity. The ratio of the amount of water vapor actually in the air compared to the amount of water vapor the air can hold at a particular temperature and pressure.

ridge. The main axis of a high-pressure system. Contrast **trough**.

rime ice. A white, granular deposit of ice that forms when supercooled water droplets come in contact with an object and freeze rapidly.

sea breeze. A coastal local wind that predominantly blows from the ocean onto land, usually during the day, and often results in cool, cloudy, and foggy coastal conditions. Also known as an *onshore breeze*. The leading edge of the breeze is termed a *sea breeze front* and can sometimes trigger thunderstorms. Contrast **land breeze**.

shear. See **wind shear**.

sheet lightning. A fairly bright lightning flash from a distant thunderstorm that illuminates a portion of its cloud.

shortwave radiation. A term most often used to describe the radiant energy emitted from the sun that is in the visible and near-ultraviolet wavelengths.

shower. Intermittent precipitation from a cumuliform cloud, often changing in intensity, but not lasting long.

sleet. A type of precipitation consisting of transparent pellets of ice 5 millimeters or less in diameter. Also called *ice pellets*.

smog. Air that has restricted visibility due to pollution; also, a photochemical haze caused by the action of solar ultraviolet radiation on air pollution. Originally smog referred to a mixture of smoke and fog.

snow. Solid precipitation in the form of hexagonal ice crystals that form at below-freezing temperatures.

snowflake. An ice crystal or aggregate of ice crystals that falls from a cloud.

snow flurries. Light showers of snow that fall intermittently, rarely causing measurable accumulations.

snow roller. Cylindrical spiral of snow shaped by wind blowing across a snow-covered surface.

snow squall. An intermittent, heavy shower of snow that greatly reduces visibility. Also known as a *snow shower*.

snow-water equivalent (SWE). A measurement of the amount of water contained within the snowpack. It can be thought of as the depth of water that would theoretically result if the entire snowpack melted instaneously.

spring freeze date. The last date in spring when the temperature falls below freezing.

squall line. Any nonfrontal line or band of active thunderstorms.

station pressure. The actual air pressure computed at an observing station.

storm surge. An abnormal rise of the sea along a shore, primarily caused by the winds of a storm, especially a hurricane.

storm track. See **jet stream**.

stratosphere. The layer of the atmosphere above the troposphere and below the mesosphere (between 6 miles and 31 miles), generally characterized by an increase in temperature with altitude.

stratus. A low, gray cloud layer with a rather uniform base. Stratus layers often bring light precipitation, most commonly drizzle.

streamline. A line on a weather map where air pressure, wind speed, and wind direction are constant.

sublimation. The process whereby ice changes directly into water vapor, skipping the liquid stage. In meteorology, sublimation can also mean the transformation of water vapor into ice. See also **deposition**.

subsidence. The slow sinking of air, usually associated with high-pressure areas.

subsidence inversion. A temperature inversion produced by the adiabatic warming of a layer of sinking air.

supercell. An often-dangerous convective storm that consists primarily of a single, long-lived, rotating updraft. This type of storm is often accompanied by strong winds and large hail, and sometimes tornadoes.

supercooled water. Water in the liquid state at a temperature well below freezing. The smaller and purer the water droplets, the more likely they can become supercooled; otherwise the water would condense on impurities (condensation nuclei).

supersaturated air. A condition that occurs in the atmosphere when the relative humidity is greater than 100 percent.

synoptic scale. The typical weather map scale that shows features such as high- and low-pressure systems and fronts over a continent-sized area. Also called the *cyclonic scale*.

temperature. The degree of hotness or coldness of a substance as measured by a thermometer. It is also a measure of the average speed or kinetic energy of the atoms and molecules in a substance.

thermal. A small rising parcel of warm air produced when the earth's surface is heated unevenly.

thermal low. An area of low atmospheric pressure near the earth's surface. It is the result of heating of the lower troposphere and the subsequent divergence of air aloft, which lowers the atmospheric pressure.

thunder. The sound caused by rapidly expanding gases along the channel of a lightning discharge. The gases expand as they are heated by the lightning.

thunder snow. A weather event that has both thunder and snow, often with the snow falling rapidly.

thunderstorm cell. The rising and falling portion of a cumulonimbus cloud that creates lightning and thunder. The rising motion is the updraft while the descending portion of the storm cloud is a precipitation downdraft. Typical thunderstorms cells last from twenty to thirty minutes. Also known as a *convective cell*.

tornado. An intense, rotating column of air that protrudes from a cumulonimbus cloud in the shape of a funnel or a rope and touches the ground. See also **funnel cloud**.

trade winds. The winds that occupy most of the tropics and blow from subtropical high-pressure areas to the equatorial low-pressure area.

transpiration. The release of water vapor to the atmosphere by plants.

tropical air mass. A warm-to-hot air mass that forms in the tropics or subtropics.

tropical depression. A mass of thunderstorms and clouds with a cyclonic wind circulation of between 23 and 39 miles per hour.

tropical disturbance. An organized mass of thunderstorms with a cyclonic wind circulation of less than 23 miles per hour.

tropical storm. Organized thunderstorms with a cyclonic wind circulation of between 40 and 73 miles per hour.

tropopause. The boundary between the troposphere and the stratosphere. Its altitude varies with latitude and season, but on average it is about 6 miles above the ground in the United States.

troposphere. The layer of the atmosphere extending from the earth's surface up to the tropopause (about 6 miles above the ground).

trough. The main axis of a low-pressure system. Contrast **ridge**.

turbulence. Any irregular or disturbed flow in the atmosphere that produces gusts and eddies.

typhoon. A hurricane that forms in the western Pacific Ocean.

ultraviolet radiation. Electromagnetic radiation (energy) with wavelengths longer than x-rays but shorter than visible light.

upslope fog. Fog that forms as moist, stable air flows upward over a topographic barrier.

upslope precipitation. Precipitation that forms due to moist, stable air gradually rising along an elevated plain. Upslope precipitation is common over the western Great Plains, especially east of the Rocky Mountains. See also **orographic precipitation**.

valley breeze. See **mountain breeze**.

valley fog. See **radiation fog**.

vapor pressure. The pressure exerted by water vapor molecules in a given volume of air.

virga. Precipitation that falls from a cloud but evaporates before reaching the ground.

visibility. The greatest distance at which an observer can see and identify prominent objects.

vortex. Air that has a whirling or circular motion that tends to form a cavity or vacuum in its center. Examples are eddies and tornadoes.

vorticity. A measurement of the rotation of a small air parcel. In the northern hemisphere, an air parcel that has a counterclockwise, or cyclonic, rotation has *positive vorticity*. Likewise, an air parcel that has a clockwise, or anticyclonic, rotation has *negative vorticity*. It is calculated based on wind shear, wind speed, and other forces.

warm front. The leading edge of a relatively warm air mass.

water balance. The comparison of actual and potential evapotranspiration with the amount of precipitation; usually recorded on a monthly basis.

waterspout. A small, weak tornado over a body of water.

water year. Also called *hydrologic year*. Precipitation measured from October 1 to September 30 in the northern hemisphere, and from July 1 to June 30 in the southern hemisphere. Climatologists often use water year versus calendar year precipitation measurements to avoid dividing the wet season into two years.

windchill factor. The cooling effect of a combination of temperature and wind, which is expressed as the loss of body heat. This describes how the temperature feels, and is not the actual air temperature. Also called *windchill index*.

wind shear. A difference in wind speed or direction between two wind currents in the atmosphere.

Index

About the Author

A self-proclaimed weather nut since childhood, **Tye W. Parzybok** earned his bachelor's degree in geography from Oregon State University. He has worked for weather services in Oregon, Utah, Washington D.C., and Colorado; helped develop a new climate atlas for the United States; and currently helps analyze extreme precipitation to help engineers design roads, culverts, dams, and other flood sensitive infrastructure. Since 1995 he has run his own meteorological consulting company, Metstat, Inc., in Windsor, Colorado.